Star Tales

Frontispiece: Orion, the giant hunter, as depicted on a chart from Johann Bayer's *Uranometria*, the first major printed star atlas, published in 1603. Bayer's atlas contained plates for each of the 48 Greek constellations catalogued in Ptolemy's *Almagest*, plus one plate for the 12 new constellations of the far-southern sky that had been invented by Dutch navigators only a few years earlier. Bayer introduced the system of labelling bright stars by Greek letters. Betelgeuse in Orion's shoulder has been assigned the letter Alpha (α). Rigel, in his foot, is labelled Beta (β). Bellatrix in Orion's other shoulder is Gamma (γ). The line of three stars forming Orion's Belt are labelled Delta (δ), Epsilon (ε), and Zeta (ζ) from right to left, while the star in Orion's sword is allocated the letter Theta (θ). Bayer's depiction of Orion was unusual, as it showed the giant from behind and facing to the left, away from Taurus the bull which is off the chart to the right. For a more conventional view of Orion, see the depiction by Johann Bode on page 133. (US Naval Observatory)

STAR TALES

Ian Ridpath

L

The Lutterworth Press
Cambridge

Published by
The Lutterworth Press
P.O. Box 60
Cambridge
CB1 2NU

www.lutterworth.com

ISBN 978 0 7188 9478 8

British Library Cataloguing in Publication Data
A record is available from the British Library

First published in 1998 by The Lutterworth Press
Expanded second edition, 2018

Typeset by Ian Ridpath
www.ianridpath.com

CONTENTS

Preface

STORYTELLING is one of the most engaging of human arts, and what greater inspiration to a storyteller's imagination than the stars of night. My interest in star tales has its roots in a series of skywatching guides that I produced in conjunction with the great Dutch celestial cartographer Wil Tirion. As I came to describe each constellation I found myself wondering about its origin and the way in which ancient people had personified it in mythology. Astronomy books did not contain satisfactory answers; they either gave no mythology at all, or they recounted stories that, I later discovered, were not true to the Greek originals. In addition, many authors seemed unaware of the true originators of several of the constellations that have been introduced since ancient Greek times. I decided, therefore, to write my own book on the history and mythology of the constellations, and a fascinating undertaking it proved to be.

My theme has been how Greek and Roman literature has shaped our perception of the constellations as we know them today – for, surprisingly enough, the constellations recognized by 21st-century science are primarily those of the ancient Greeks, interspersed with more recent additions. To this end, I have gone back to original Greek and Latin sources wherever possible; for a list of sources and references see pages 211–13. While I have attempted to recount the main variants of each myth, and to identify the writer concerned, it should be realized that there is no such thing as a 'correct' myth; for some stories, there are almost as many different versions as there are mythologists.

I should also make clear what this book is not about: I have not tried to compare the Greek and Roman constellations with the constellations that were imagined by other cultures such as the Chinese, Egyptian, or Hindu. Such a diversion would have taken me too far from my intended task. Neither have I delved too far into the confusing morass of speculation about when and where the very first constellations originated. Indeed we may never be able to provide convincing answers to those questions from the fragmentary information available. In this second edition of the book, though, I have provided additional background about the origin and history of the constellations that have been introduced by astronomers since ancient Greek times including 24 now-obsolete figures that gained at least some degree of currency.

Since ancient astronomers regarded each constellation as embodying a picture of a mythological character or an animal, rather than as simply an area of sky as defined by today's surveyor-astronomers, it seemed natural to illustrate each constellation with a picture from an old star map. Most of the illustrations I have chosen come from either Johann Bode's *Uranographia* of 1801 or John Flamsteed's *Atlas Coelestis* of 1729, probably the two greatest star atlases of all time. These celestial charts are works of art in themselves, and are among the most elegant treasures bequeathed to us by astronomers of the past. The constellations give us a very real link with the most ancient civilizations. It is a heritage that we can share whenever we look at the night sky.

Ian Ridpath
Brentford, 2018

CHAPTER ONE

Stars and storytellers

EVERY NIGHT a pageant of Greek mythology circles overhead. Perseus flies to the rescue of Andromeda, Orion faces the charge of the snorting bull, Boötes herds the bears around the pole, and the ship of the Argonauts sails in search of the golden fleece. These legends, along with many others, are depicted in the star patterns that astronomers term constellations.

Constellations are the invention of human imagination, not of nature. They are an expression of the human desire to impress its own order upon the apparent chaos of the night sky. For navigators beyond sight of land or for travellers in the trackless desert who wanted signposts, for farmers who wanted a calendar and for shepherds who wanted a nightly clock, the division of the sky into recognizable star groupings had practical purposes. But perhaps the earliest motivation was to humanize the forbidding blackness of night.

Newcomers to astronomy are soon disappointed to find that the great majority of constellations bear little, if any, resemblance to the figures whose names they carry; but to expect such a resemblance is to misunderstand their true meaning. The constellation figures are not intended to be taken literally. Rather, they are symbolic, a celestial allegory. The night sky was a screen on which human imagination could project the deeds and personifications of deities, sacred animals, and moral tales. It was a picture book in the days before writing.

Each evening the stars emerge like magic spirits as the Sun descends to its nocturnal lair. Modern science has told us that those twinkling points scattered across the sky are actually glowing balls of gas similar to our own Sun, immensely far away. A star's brightness in the night sky is a combination of its own power output and its distance from us. So far apart are the stars that light from even the nearest of them takes many years to reach us. The human eye, detecting the faint spark from star fires, is seeing across unimaginable gulfs of both space and time.

Such facts were unknown to the ancient Greeks and their predecessors, to whom we owe the constellation patterns that we recognize today. They were not aware that, with a few exceptions, the stars of a constellation have no connection with each other, but lie at widely differing distances. Chance alone has given

Figure 1: The 48 constellations of the Greek astronomer Ptolemy, illustrated on a pair of woodcuts made by the German artist Albrecht Dürer in 1515, one showing the northern sky (left) and the other the southern sky (right). The figures are depicted from the rear, as

us such familiar shapes as the 'W' of Cassiopeia, the square of Pegasus, the sickle of Leo, or the Southern Cross.

The constellation system that we use today has grown from a list of 48 constellations published around AD 150 by the Greek scientist Ptolemy in an influential book called the *Almagest* (see Table 1, page 15). Since then various astronomers have added another 40 constellations, filling the gaps between Ptolemy's figures and populating the region around the south celestial pole that was below the horizon of the Greeks. The result is a total of 88 contiguous constellations that all astronomers accept by international agreement. The tales of these constellations are told in this book – along with two dozen others that fell by the wayside.

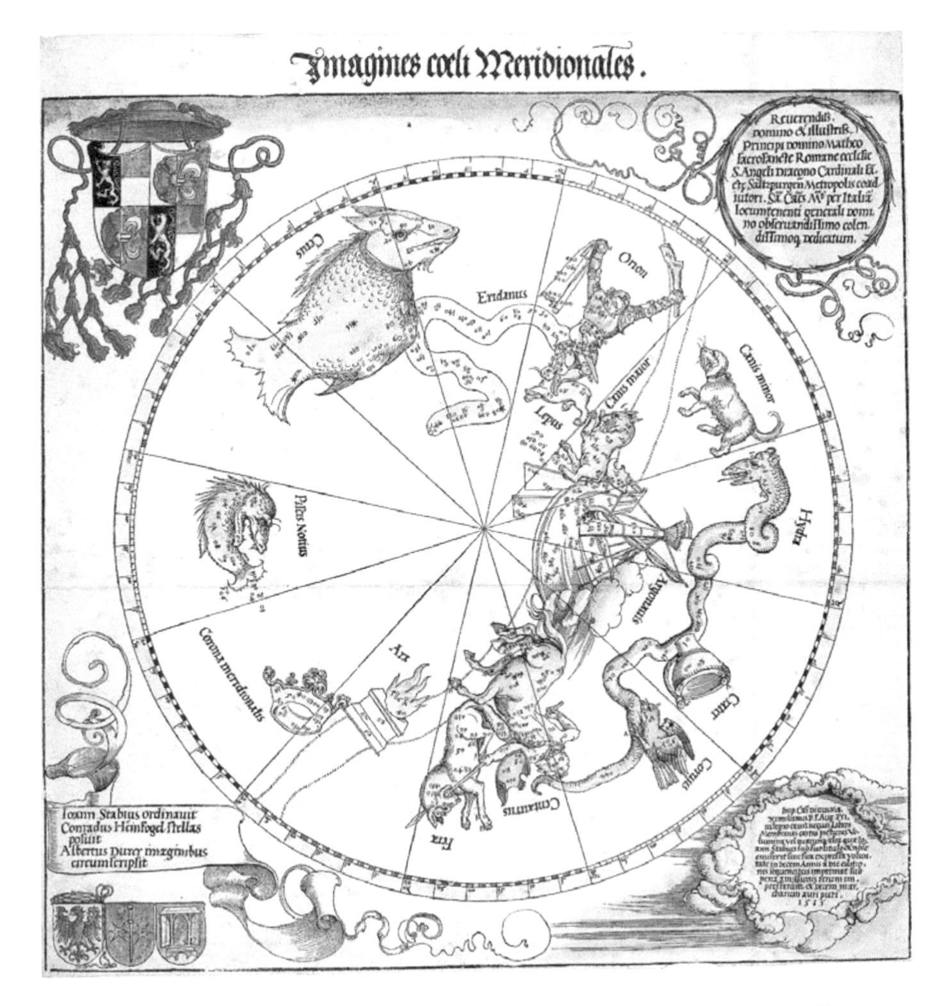

on a celestial globe. Note the large blank area of the southern sky that was below the horizon to the people who invented the constellations. The size of this blank zone is a clue to the latitude at which the constellation inventors lived. (Sotheby's)

Ptolemy did not invent the constellations that he listed. They are much older than his era, although exactly when and where they were invented is lost in the mists of time. The early Greek writers Homer and Hesiod (*c.*700 BC) mentioned only a few star groups, such as the Great Bear, Orion, and the Pleiades star cluster (the Pleiades was then regarded as a separate constellation rather than being incorporated in Taurus as it is today).

The major developments evidently took place farther east, around the Tigris and Euphrates rivers in what is now Iraq. There lived the Babylonians, who at the time of Homer and Hesiod had a well-established system of constellations of the zodiac, the strip of sky traversed by the Sun, Moon, and planets. We know this from a star list written in cuneiform on a clay tablet dated to 687 BC.

Scholars call this list the MUL.APIN series, from the first name recorded on the tablet. The Babylonian constellations had many similarities with those we know today, but they are not all identical. From other texts, historians have established that the constellations known to the Babylonians actually originated much earlier, with their ancestors the Sumerians before 2000 BC.

1.1. *Eudoxus, Aratus, and the Phaenomena*

If the Greeks of Homer and Hesiod's day knew of the Babylonian zodiac they did not write about it. The first clear evidence we have for an extensive set of Greek constellations comes from the astronomer Eudoxus (*c.*390–*c.*340 BC). Eudoxus reputedly learned the constellations from priests in Egypt and introduced them to Greece, which makes his contribution to astronomy highly significant. He published descriptions of the constellations in two works called *Enoptron* (Mirror) and *Phaenomena* (Appearances). Both these works are lost, but the *Phaenomena* lives on in a poem of the same name by another Greek, Aratus (*c.*315–*c.*245 BC). Aratus's *Phaenomena* gives us a complete guide to the constellations known to the ancient Greeks; hence he is a major figure in our study of constellation lore.

Aratus was born at Soli in Cilicia, on the southern coast of what is now Turkey. He studied in Athens before going to the court of King Antigonus of Macedonia in northern Greece. There, at the king's request, he produced his poetic version of the *Phaenomena* of Eudoxus around 275 BC. In the *Phaenomena* Aratus identified 48 constellations, although they are not the exact same as Ptolemy's 48. Aratus included the Water (now regarded as part of Aquarius) and the Pleiades, while Corona Australis was only alluded to as an anonymous ring of stars beneath the feet of Sagittarius; evidently it had not yet become a separate constellation. Neither does he mention Equuleus, which first appeared in the *Almagest* four centuries later.

Aratus also named six individual stars: Arcturus (Ἀρκτοῦρος); Capella, which he called Aix (Αἴξ); Sirius (Σείριος); Procyon (Προκύων), which formed a constellation on its own; Spica, which he called Stachys (Στάχυς); and Vindemiatrix, which he called Protrygeter (Προτρυγητήρ). This last star is a surprise, since it is so much fainter than the others, but the Greeks used it as a calendar star because its rising at dawn in August marked the start of the grape harvest.

Neither the Greeks nor the Egyptians actually invented the constellations that are described in the *Phaenomena*. The evidence for that statement lies not just in written records, but in the sky itself.

1.2. *Identifying the constellation inventors*

It is not too difficult to work out roughly where the constellations known to Eudoxus and Aratus were invented. The clue is that Aratus described no constellations around the south celestial pole, for the reason that this area of sky was permanently below the horizon of the constellation makers. From the extent of the constellation-free zone we can conclude that the constellation makers must have lived at a latitude of around 35°–36° north – that is, south of Greece but north of Egypt.

A second clue comes from the fact that the constellation-free zone is centred not on the south celestial pole at the time of Aratus but on its position many centuries earlier. The position of the celestial pole changes slowly with time because of a wobble of the Earth on its axis, an effect known as precession, and in principle this effect can be used to deduce the date of any set of star positions.

Because of the uncertainties involved, however, attempts to date the constellations as described by Aratus have produced a wide range of results. Derived values extend back to nearly 3000 BC, with a majority preference for somewhere around 2000 BC. A more comprehensive analysis by Bradley Schaefer of Louisiana State University published in 2004 concluded that Aratus's descriptions correspond to the sky as it appeared close to 1130 BC.

At present, the best we can say is that the constellations known to Eudoxus and Aratus were probably invented in the second millennium BC by people who lived just south of latitude 36° north. This date is too early for the Greeks and the latitude is too far south; Egyptian civilization is sufficiently old, but the required latitude is well north of them. The time and the place, though, ideally match the Babylonians and their Sumerian ancestors who lived in the area we know as Mesopotamia and who, as we have already seen, had a well-developed knowledge of astronomy by 2000 BC. Hence two independent lines of evidence point to the Babylonians and Sumerians as the originators of our constellation system.

But why had the constellation system introduced by Eudoxus not been updated by its makers to take account of the changing position of the celestial pole? As we have seen, the constellations introduced by Eudoxus and described by Aratus in the *Phaenomena* refer to the position of the celestial pole around a thousand years earlier. By the time of Aratus, the shift in position of the celestial pole meant that certain stars mentioned in the *Phaenomena* were now permanently below the horizon from latitude 36° north, while others not mentioned by Aratus had by then come into view. Oddly, Eudoxus himself seems not to have been bothered by these anomalies, if he even noticed them; but the great Greek astronomer Hipparchus (*fl.*146–127 BC) recognized the differences and was understandably critical.

Archie Roy of Glasgow University has argued that the Babylonian constellations reached Egypt (and hence Eudoxus) via some other civilization; he proposes that they were the Minoans who lived on Crete and the surrounding islands off the coast of Greece, including Thera (also known as Santorini). Crete lies between 35° and 36° north, which is the right latitude, and the Minoan empire was expanding between 3000 and 2000 BC, which is the right date. What's more, the Minoans were in contact with the Babylonians through Syria from an early stage. Hence they must have been familiar with the old Babylonian constellations, and they could well have adapted the Babylonian star groups into a practical system for navigation.

But the Minoan civilization was wiped out around 1700 BC by the explosive eruption of a volcano on the island of Thera about 120 km north of Crete. It was one of the greatest natural catastrophes in the history of civilization, the probable origin of the legend of Atlantis. Professor Roy supposes that Minoan refugees brought their knowledge of the stars to Egypt after the eruption, where it was eventually encountered by Eudoxus in unchanged form over a thousand years later.

The thesis is an attractive one, for it is easy to imagine the Minoans utilizing the Babylonian constellations in this way. In addition, many star myths are centred on Crete. However, it must be admitted that there is no direct evidence, such as wall paintings or star lists like those of the Babylonians, to demonstrate any Minoan interest in astronomy. So, for now, the theory that the Minoans were middlemen to our constellation system remains nothing more than an appealing speculation.

1.3. *The mythographers*

The *Phaenomena* of Aratus was immensely popular and became an essential part of the culture not just during the Greek era but for many centuries afterwards. It was translated several times into Latin, often with extensive commentaries by its translators and editors, and medieval versions were highly illustrated. For our purposes the most useful version is a Latin adaptation of Aratus attributed to Germanicus Caesar (15 BC–AD 19), which has more information about the identification of certain constellations than Aratus's original. According to the classicist David B. Gain this Latin version of the *Phaenomena* could have been written either by Germanicus himself or by his uncle (and adoptive father) Tiberius Caesar, but in this book I refer to the author simply as Germanicus.

After Aratus, the next landmark in our study of Greek constellation lore is Eratosthenes (*c*.276–*c*.194 BC), to whom an essay called the *Catasterisms* is attributed. Eratosthenes was a Greek scientist and writer who worked in Alexandria at the mouth of the Nile. The *Catasterisms* gives the mythology of 42 separate constellations (the Pleiades cluster is treated individually), with a listing of the main stars in each figure. The version of the *Catasterisms* that survives is only a summary of the original, made at some unknown date, and it is not even certain that the original was written by the real Eratosthenes; hence the author of the *Catasterisms* is usually referred to as pseudo-Eratosthenes. The antiquity of his sources is certain, though, because he quotes in places from a long-lost work on astronomy by Hesiod (*c*.700 BC).

Another influential source of constellation mythology is a book called *Poetic Astronomy* by a Roman author named Hyginus, apparently written in the second century AD. We do not know who Hyginus was, not even his full name – he was evidently not C. Julius Hyginus, a Roman writer of the first century BC. *Poetic Astronomy* is based on the constellations listed by Eratosthenes (Hyginus differs only by including the Pleiades under Taurus), but it contains many additional stories. Hyginus also wrote a compendium of general mythology called the *Fabulae*. In medieval and Renaissance times many illustrated versions of Hyginus's writings on astronomy were produced, the most famous of which was published by the German printer Erhard Ratdolt in Venice in 1482.

Marcus Manilius, a Roman author of whom virtually nothing is known, wrote a book called *Astronomica* around AD 15, clearly influenced by the *Phaenomena* of Aratus. Manilius's book deals mostly with astrology rather than astronomy

PTOLEMY'S ALMAGEST

The source of our modern constellations

In the *Almagest* Ptolemy listed 1,028 objects forming the classical 48 constellations. Three stars were deliberately entered twice, since Ptolemy regarded them as being shared between constellations: these were the stars we know as Alpha Piscis Austrini (Fomalhaut), Beta Tauri (Elnath), and Nu Boötis. Another three entries in Ptolemy's catalogue are actually not stars at all but star clusters: the Double Cluster in Perseus; M44 (Praesepe) in Cancer; and the globular cluster Omega Centauri. Hence it is usually said that the *Almagest* contains 1,022 stars. However, the stars 18 and 20 in Cetus are now thought to be duplicates of Cetus 17 and 19, so the total number of separate stars in the *Almagest* is actually 1,020. The number of stars Ptolemy catalogued in each constellation ranged from a mere two in Canis Minor to 45 in Argo Navis.

At the end of some constellations Ptolemy listed what he called ἀμόρφωτοι, i.e. *amorphotoi* – 'unformed' stars (*informatae* in Latin) that lay outside the recognized constellation pattern. Most of these unformed stars have since been absorbed into the relevant constellation or a neighbour, although in some cases later astronomers incorporated the orphan stars into new constellations.

Ptolemy's catalogue was not fully superseded until the end of the 16th century when the Danish astronomer Tycho Brahe produced a thousand-star catalogue that was ten times more accurate, heralding a new era of star surveying (see page 18).

Table 1

The 48 constellations listed by the Greek astronomer Ptolemy in the *Almagest*, second century AD (modern Latin names)

Andromeda	Cetus	Lyra
Aquarius	Corona Australis	Ophiuchus
Aquila	Corona Borealis	Orion
Ara	Corvus	Pegasus
Argo Navis (*now subdivided into Carina, Puppis, and Vela*)	Crater	Perseus
	Cygnus	Pisces
	Delphinus	Piscis Austrinus
Aries	Draco	Sagitta
Auriga	Equuleus	Sagittarius
Boötes	Eridanus	Scorpius
Cancer	Gemini	Serpens
Canis Major	Hercules	Taurus
Canis Minor	Hydra	Triangulum
Capricornus	Leo	Ursa Major
Cassiopeia	Lepus	Ursa Minor
Centaurus	Libra	Virgo
Cepheus	Lupus	

but it contains numerous insights into constellation lore and I have quoted him a number of times throughout this book.

The names of three other mythologists appear frequently on the following pages, and although they are not astronomers they must be introduced before we return to the history of the constellations. Foremost among them is the Roman poet Ovid (43 BC–AD 17), who recounts many famous myths in his books the *Metamorphoses*, which deals with transformations of all kinds, and the *Fasti*, a treatise on the Roman calendar. Apollodorus was a Greek who compiled an almost encyclopedic summary of myths known as the *Library* some time in the late first century BC or the first century AD. Finally there is the Greek writer Apollonius Rhodius (Apollonius of Rhodes) whose *Argonautica*, an epic poem on the voyage of Jason and the Argonauts composed in the third century BC, includes much mythological information. These are the main sources for the stories in this book.

1.4. *Ptolemy's 48 constellations*

Greek astronomy reached its pinnacle with Ptolemy (*c.* 100–*c.* 178) who worked in Alexandria, Egypt. Around AD 150, Ptolemy produced a summary of Greek astronomical knowledge usually known by its later Arabic title of the *Almagest* meaning 'the greatest'. At its heart was a catalogue of over a thousand stars arranged into 48 constellations (see Table 1, previous page), with estimates of their brightness, based largely on the observations of the Greek astronomer Hipparchus three centuries earlier.

Ptolemy did not identify the stars in his catalogue by means of Greek letters as astronomers do today, but described their position within each constellation figure. For instance, the star in Taurus which Ptolemy referred to as 'the reddish one on the southern eye' is known today as Aldebaran. At times, this system became cumbersome: 'The northernmost of the two stars close together over the little shield in the poop' is how Ptolemy struggled to identify a star (now called Xi Puppis) in the constellation of Argo Navis.

The tradition of describing stars by their positions within a constellation had already been established by Eratosthenes and Hipparchus. Clearly, the Greeks regarded the constellations not merely as assemblages of stars but as true pictures in the sky. Identification would have been easier if they had given the stars individual names, but Ptolemy added only four stars to those named by Aratus four centuries earlier: Altair which Ptolemy called Aetos (Ἀετός), meaning eagle; Antares (Ἀντάρης); Regulus which he called Basiliskos (Βασιλίσκος); and Vega which he called Lyra (Λύρα), the same name as its constellation.

It would be difficult to overemphasize the influence of Ptolemy on astronomy; the constellation system we use today is essentially Ptolemy's, modified and extended. Mapmakers in Europe and Arabia used his constellation figures for over 1,500 years, witness this passage from the preface to the *Atlas Coelestis* by the first Astronomer Royal of England, John Flamsteed, published in 1729:

> From Ptolemy's time to ours the names that he made use of have been continued by the ingenious and learned men of all nations; the Arabians always used his forms and names of the constellations; the old Latin

catalogues of the fixed stars use the same; Copernicus's catalogue and Tycho Brahe's use the same; so do the catalogues published in the German, Italian, Spanish, Portuguese, French and English languages. All the observations of the ancients and moderns make use of Ptolemy's forms of the constellations and names of the stars so that there is a necessity of adhering to them, that we may not render the old observations unintelligible by altering or departing from them.

1.5. *Arabic influences*

After Ptolemy, Greek astronomy went into permanent eclipse. By the eighth century AD the centre of astronomy had moved east from Alexandria to Baghdad where Ptolemy's work was translated into Arabic and received the name *Almagest* by which we still know it. ʿAbd al-Raḥmān al-Ṣūfī (AD 903–986), one of the greatest Arabic astronomers (also known by the Latinized name of Azophi), produced his own version of Ptolemy's star catalogue called *Kitāb al-Kawākib al-Thābita* (*Book of the Fixed Stars*) in which he recorded many Arabic star names.

Bedouin Arabs had their own names for various bright stars, an example being Aldebaran which we have inherited from them. They also had a very different star lore from the Greeks, and commonly regarded single stars as representing animals or people. For example, the stars we know as Alpha and Beta Ophiuchi were regarded by the Arabs as a shepherd and his dog, while neighbouring stars made up the outlines of a field with sheep. Elsewhere could be found camels, gazelles, ostriches, and a family of hyenas.

Some of the Arabic names were already so many centuries old by the time of al-Ṣūfī that their meanings were lost even to him and his contemporaries, and they remain unknown today. Other star names used by al-Ṣūfī and his compatriots were direct translations of Ptolemy's descriptions. For example, the star name Fomalhaut comes from the Arabic meaning 'mouth of the southern fish', which is where Ptolemy had described it in the *Almagest.*

Another rich source of Arabic star names were astrolabes, star-finding devices like a flattened celestial sphere invented by the Greeks but developed to the height of sophistication by the Arabs. Each astrolabe had a rotating disk with decorative pointers that indicated the positions of various bright stars, the names of which were engraved on the pointer to assist identification.

From the tenth century onwards the translated works of Ptolemy were introduced into Europe by Islamic Arab incursions. There they were retranslated from Arabic into Latin, the scientific language of the day. The Spanish city of Toledo, in particular, is said to have become a veritable translation factory during the 12th century and scholars flocked there from all over western Europe to study the marvellous new works – not just on astronomy but mathematics and all other branches of science. It is through this roundabout route of old Greek writings being transmitted through Arabic hands and then translated back into Latin in Europe in the middle ages that we have ended up with a polyglot system of Greek constellations with Latin names containing stars with a mixture of Arabic and Greek titles.

1.6. *Extending Ptolemy's 48*

Although Arab astronomers increased the number of star names, the number of constellations remained unchanged. The first extension of Ptolemy's 48 was made in 1536 on a celestial globe by the German mathematician and cartographer Caspar Vopel (1511–61) who depicted Antinous and Coma Berenices as separate constellations; Ptolemy had mentioned these star groups in the *Almagest*, but only as subdivisions of Aquila and Leo respectively. Vopel's lead was followed in 1551 on a celestial globe by the great Dutch cartographer Gerardus Mercator (1512–94). The Danish astronomer Tycho Brahe (1546–1601) listed Antinous and Coma Berenices separately in his influential star catalogue of 1602 (see box), ensuring their widespread adoption. Coma Berenices is still a recognized constellation, but Antinous has since been remerged with Aquila.

By now the European age of exploration was well under way and navigator–astronomers turned their attentions to the hitherto uncharted regions of the sky in the southern hemisphere which had been below the horizon for the ancient Greeks. Three names stand out from this era. The first is Petrus Plancius (1552–1622), a Dutch theologian and cartographer; his name is the Latinized form of Pieter Platevoet – literally, Peter Flatfoot. The other two were the Dutch navigators Pieter Dirkszoon Keyser (*c.*1540–96), also known as Petrus Theodorus or Peter Theodore, and Frederick de Houtman (1571–1627). Surprisingly, all three are little-known today despite their lasting contributions.

TYCHO BRAHE'S STAR CATALOGUE
First true successor to the *Almagest*

During the last two decades of the 16th century the Danish astronomer Tycho Brahe (1546–1601), usually known simply as Tycho, exercised his exceptional ingenuity and energy to produce the first major star catalogue since the *Almagest* over 1,400 years earlier. Tycho's observatory, called Stjerneborg, was on the Danish island of Hven. His largest instrument, a wall-mounted quadrant, was in his adjacent castle called Uraniborg. Tycho's obsessive attention to detail resulted in a tenfold improvement in positional accuracy over his predecessors. Working solely with naked-eye instruments – the telescope had not yet been invented – he set new standards in celestial surveying and provided reliable data for constellation chartmakers.

Because he worked at latitude 55.9°, some 25° farther north than Ptolemy, Tycho was unable to observe the more southerly stars in the *Almagest*, but he observed additional stars in most of the other Ptolemaic constellations. An abridged version of Tycho's catalogue, containing 777 of the most accurately determined star positions divided into 45 constellations, was printed in 1602, the year after his death. This formed the basis of the first great celestial atlas, Johann Bayer's *Uranometria*, published in 1603 (see Section 2.4, page 32). Tycho's full catalogue of 1,004 entries was edited and published in 1627 by his former assistant, the German mathematician Johannes Kepler (1571–1630).

Table 2

Twelve southern constellations introduced 1596–1603 from the observations of Pieter Dirkszoon Keyser and Frederick de Houtman

Apus	Hydrus	Phoenix
Chamaeleon	Indus	Triangulum Australe
Dorado	Musca	Tucana
Grus	Pavo	Volans

1.7. *Scouting the southern sky*

Plancius instructed Keyser to make observations to fill in the constellation-free zone around the south celestial pole. Keyser was chief pilot on the *Hollandia* and later on the *Mauritius*, two of the fleet of four ships that left the Netherlands in 1595 on the first Dutch trading expedition to the East Indies. The outbound expedition spent several months anchored at Madagascar and it was there that Keyser made most of his observations. The Dutch historian and geographer Paul Merula (1558–1607) wrote in *Cosmographiae Generalis* (1605) that Keyser observed from the crow's nest using an instrument given to him by Plancius. This instrument was probably either a cross-staff or a universal astrolabe (sometimes known as an *astrolabium catholicum*), as this was still the pre-telescopic era.

Keyser died in September 1596 while the fleet was at Bantam (now Banten, near the modern Serang in western Java). His observations were delivered to Plancius when the fleet returned to Holland the following year. Regrettably, little else seems to be known about Keyser's life and accomplishments, but he left his mark indelibly on the sky.

Keyser's stars, divided into 12 newly invented constellations, first appeared on a globe by Plancius in 1598, and again two years later on a globe by the Dutch cartographer Jodocus Hondius (1563–1612). The acceptance of these new constellations was assured when Johann Bayer, a German astronomer, included them in his *Uranometria* of 1603, the leading star atlas of its day (see Section 2.4, page 32). Keyser's observations were eventually published in tabular form by Johannes Kepler in the *Rudolphine Tables* of 1627.

Unfortunately, Keyser's original manuscript is long lost and so we do not know whether he sorted his stars into the 12 new southern constellations himself or whether that was done later by someone else. If it was not Keyser, then a plausible candidate for the inventor of the dozen southern constellations is Frederick de Houtman (see box overleaf), younger brother of the commander of the Dutch fleet to the East Indies, Cornelis de Houtman (1565–99). Frederick was also a member of the crew and made celestial observations of his own, independently of Keyser.

After Keyser died, Frederick de Houtman would have had access to his records and might well have taken custody of them on the long voyage home. We can easily imagine him whiling away the time at sea by collating the joint observations, grouping them into constellations representing the wondrous things they had seen, and planning a more extensive observing campaign for his next voyage south, which was not long coming.

1.8. *A second voyage south*

The de Houtman brothers departed on a second voyage to the East Indies in 1598. During this voyage Cornelis was killed and Frederick was imprisoned for two years by the Sultan of Atjeh in northern Sumatra. Frederick made good use of his time as a captive by studying the local Malay language and making astronomical observations. In 1603, following his return to Holland, Frederick published his observations as an appendix to a Malay and Madagascan dictionary that he had compiled – one of the most unlikely pieces of astronomical publishing in history. In the Introduction he wrote: 'Also added [are] the declination of several fixed stars which during the first voyage I have observed around the south pole; and during the second [voyage], in the island of Sumatra, improved upon with greater diligence, and increased in number.'

De Houtman increased Keyser's measured star positions to 303, although 107 of these were stars already known to Ptolemy, according to a study of

FREDERICK DE HOUTMAN'S CATALOGUE

Explorer of the southern sky

The oldest surviving catalogue of the southern stars was made by the Dutch seafarer Frederick de Houtman (1571–1627) from Sumatra and published in Amsterdam in 1603. De Houtman made some observations of the southern stars on his first voyage to the East Indies in 1595–97 and revised and increased the number on his second voyage in 1598–1602.

De Houtman listed 304 stars (one without coordinates), 111 of them lying in the 12 new southern constellations which had been invented during or just after his first voyage. The bulk of his catalogue, though, was devoted to filling out the existing Ptolemaic figures – in particular, he gave positions for 56 stars in Argo Navis and 48 in Centaurus, of which 52 were new. He listed Crux as a separate constellation ('De Cruzero') for the first time.

The 12 new southern constellations as listed by de Houtman in Dutch, with their present-day names in brackets, are:

Den voghel Fenicx (Phoenix); De Waterslang (Hydrus); Den Dorado (Dorado); De Vlieghe (Musca); De vlieghende Visch (Volans); Het Chameljoen (Chamaeleon); Den Zuyder Trianghel (Triangulum Australe); De Paradijs Voghel (Apus); De Pauww (Pavo); De Indiaen (Indus); Den Reygher – literally 'the heron' (Grus); Den Indiaenschen Exster, op Indies Lang ghenaemt – literally 'the Indian magpie, named Lang in the Indies' (Tucana). In addition, he listed stars in the pre-existing constellations of Ara, Argo Navis, Centaurus, Corona Australis, Crux, Lupus, Columba (which he called De Duyve met den Olijftak – literally 'The dove with olive branch'), the tail of Scorpius, and southern Eridanus, which he termed 'den Nyli', the Nile.

THE CONSTELLATIONS OF PETRUS PLANCIUS
Adding to Ptolemy's 48

Petrus Plancius (1552–1622), Dutch cartographer and constellation inventor, left no written records so what we know of his role in the development of our system of constellations is based on examination of his surviving maps and globes. His first foray into celestial mapping came on a terrestrial map of 1592 which contained small insets showing the northern and southern sky. Among the constellations were two inventions of his own: Columba, the dove, and Polophylax, a pole-watcher, intended as the southern equivalent of Boötes (which the Greeks termed Arctophylax, i.e. bear-watcher). Columba, formed from stars listed in Ptolemy's *Almagest* south of Canis Major, became established. Polophylax, based on sketchy information about the southern stars and positioned between Piscis Austrinus and the southern pole in an area now occupied by Grus and Tucana, did not.

In 1598 Plancius produced a globe in conjunction with fellow Dutchman Jodocus Hondius (1563–1612) that was a landmark in constellation history. For the first time 12 new southern constellations were shown (see Table 2), based on the observations of Pieter Dirkszoon Keyser which had been brought back from the East Indies in 1597 after Keyser's death.

A later Plancius globe of 1612 introduced Camelopardalis and Monoceros, along with others in both hemispheres that never gained acceptance: Jordan, Tigris, Apes, Gallus, Cancer Minor, and Sagitta Australe. All but the last two first appeared in print (as distinct from on a globe) on the charts in Jacob Bartsch's book *Usus Astronomicus Planisphaerii Stellati* (Astronomical Use of the Stellar Planisphere) published in 1624, which led some to wrongly attribute their formation to Bartsch himself.

the catalogue by the English astronomer Edward Ball Knobel. Nowhere did de Houtman give Keyser credit for his priority – in fact, relations between the two men seem to have broken down during their voyage together, despite their common interest in the sky.

De Houtman's catalogue of southern stars, divided into the same 12 constellations as shown on the globes of Plancius and Hondius, was used by the Dutch cartographer Willem Janszoon Blaeu (1571–1638) for his celestial globes from 1603 onwards. Keyser and de Houtman are now credited jointly with the invention of these 12 southern constellations, which are still recognized today (see Table 2, page 19).

The Dutch historian Elly Dekker has argued that the true credit for dividing the newly observed stars into 12 constellations is actually due to Petrus Plancius, after he received Keyser's observations in 1597. Whatever the case, Plancius invented some other constellations that are indubitably his own, among them Columba, the dove, which he formed from nine stars that Ptolemy had listed as surrounding Canis Major. He also invented the unlikely sounding Monoceros,

THE STAR CATALOGUE AND ATLAS OF JOHANNES HEVELIUS

Johannes Hevelius (1611–87), compiler of the last major star catalogue to be made with naked-eye instruments, was a wealthy brewer from Danzig, now Gdańsk, Poland. In the 1640s he set out to enlarge and improve upon the star catalogue of Tycho Brahe. Hevelius observed from a platform over the roof of his house with naked-eye instruments such as a quadrant and sextant, assisted from 1663 by his second wife, Elizabeth (*c.*1646/7–*c.*1693). In 1679 a fire destroyed much of the building but his precious catalogue was saved. This, along with his star atlas, was in the process of being printed when Hevelius died in 1687. Elizabeth supervised its final publication in 1690.

Hevelius's master work came in three parts: an introduction called *Prodromus Astronomiae*, which included descriptions of the ten new constellations he had invented; the catalogue of 1,564 stars, called *Catalogus Stellarum Fixarum*; and the star atlas, *Firmamentum Sobiescianum*. Seven of the constellations introduced by Hevelius are still accepted and are listed in Table 3. Of these seven, Scutum had already been published in 1684 to honour the King of Poland who had helped Hevelius with rebuilding his observatory after the destructive fire. The remainder had been invented by 1687, the date on the printed catalogue, although they were not published until 1690. Three other Hevelius constellations shown on his charts – Cerberus, Mons Maenalus, and Triangulum Minus – were later dropped by other astronomers.

Table 3

Seven constellations introduced by Johannes Hevelius in his star catalogue of 1687

Canes Venatici	Leo Minor	Scutum	Vulpecula
Lacerta	Lynx	Sextans	

the unicorn, and Camelopardalis, the giraffe, from faint stars uncharted by Ptolemy. These three Plancius constellations are still accepted by astronomers, but his other inventions fell by the wayside.

1.9. *Filling the remaining gaps*

As the accuracy of astronomical observations improved and fainter stars were charted, the opportunities grew for innovators to introduce new constellations even among the area of sky known to the ancient Greeks. Ten more constellations were introduced later in the 17th century by the Polish astronomer Johannes Hevelius (1611–87), filling the remaining gaps in the northern sky. These were listed in his star catalogue dated 1687 and depicted on his accompanying star atlas called *Firmamentum Sobiescianum*, both published posthumously in 1690. Oddly, Hevelius insisted on measuring star positions with the naked eye even though he possessed telescopes for observing the Moon and planets; many of his constellations were deliberately faint as though he was boasting of the power of his eyesight. Of Hevelius's inventions, seven are still accepted by astronomers (see Table 3). The rejected three were Cerberus, Mons Maenalus, and Triangulum Minus.

Although the northern constellations were now complete, there were still gaps in the southern sky. These were filled by the French astronomer Nicolas Louis de Lacaille (1713–62) who sailed to South Africa in 1750. There he set up a small observatory at the Cape of Good Hope (not yet known as Cape

NICOLAS LOUIS DE LACAILLE AT THE CAPE

Fourteen more southern constellations

From August 1751 to July 1752 the French astronomer Nicolas Louis de Lacaille (1713–62) observed the southern skies from the rear of a house near Table Bay at the Cape of Good Hope, using a telescope of a mere 13.5 mm (half an inch) aperture mounted on a 3-ft quadrant. With this basic equipment he diligently compiled accurate observations of some 9,800 stars between the Tropic of Capricorn and the south celestial pole. He marked 1,930 naked-eye stars on a planisphere which he presented to the French Academy of Sciences in 1754. Lacaille's planisphere included the 14 new constellations he invented to accommodate the otherwise unadopted stars he had catalogued.

The names of his constellations in French, with modern equivalents, were: l'Atelier du Sculpteur (Sculptor); la Boussole (Pyxis); les Burins (Caelum); le Chevalet et la Palette (Pictor); le Compas (Circinus); l'Equerre et la Regle (Norma); le Fourneau (Fornax); l'Horloge (Horologium); la Machine Pneumatique (Antlia); le Microscope (Microscopium); Montagne de la Table (Mensa); l'Octans de Reflexion (Octans); le Reticule Romboide (Reticulum); le Telescope (Telescopium).

Lacaille's final catalogue, *Coelum Australe Stelliferum*, containing 1,942 entries was published posthumously in 1763. It included the same planisphere as before but this time with the constellation names in Latin rather than French and the stars identified with Greek and Roman letters. In both his initial and final catalogues Lacaille divided the stars of Argo Navis into three parts – the keel, the stern, and the sails – but his charts still showed it as a single figure.

Table 4

Fourteen constellations introduced by Nicolas Louis de Lacaille in 1754

Antlia	Horologium	Octans	Sculptor
Caelum	Mensa	Pictor	Telescopium
Circinus	Microscopium	Pyxis	
Fornax	Norma	Reticulum	

Town) under the famous Table Mountain, which impressed him so much that he later named a constellation after it, Mensa. At the Cape from August 1751 to July 1752 Lacaille observed the positions of nearly 10,000 stars, an astounding total in the short time.

On his return to France in 1754, Lacaille presented a map of the southern skies to the French Royal Academy of Sciences which included 14 new constellations of his own invention (see Table 4). An engraved version of the map was published in the Academy's *Mémoires* in 1756 along with a preliminary catalogue and Lacaille's new constellations were rapidly accepted by other astronomers.

Whereas Keyser and de Houtman had mostly named their constellations after exotic animals, Lacaille commemorated instruments of science and the arts, with the exception of Mensa, named after the Table Mountain under which he had carried out his observations. His full catalogue, and a revised map with the names of his new constellations in Latin, was published posthumously

Figure 2: Official boundaries to the constellations were fixed in 1928 by a Belgian astronomer, Eugène Delporte, acting on behalf of the International Astronomical Union. Here is his chart for part of the northern sky, including Cassiopeia and Andromeda, from *Délimitation Scientifique des Constellations* (1930). The constellation boundaries follow circles of right ascension (the equivalent of longitude in the sky) and parallels of declination (the celestial equivalent of latitude). In this new and more scientific depiction of the sky, the old constellation figures have gone for good. (Author's collection)

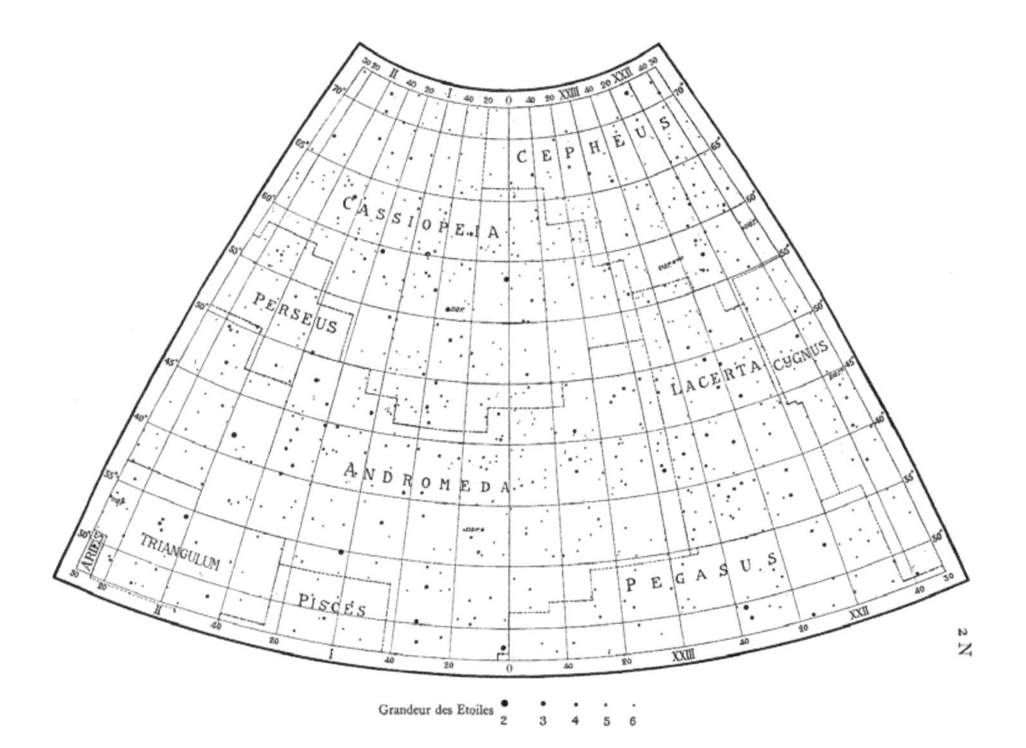

in 1763 under the title *Coelum Australe Stelliferum*. In his catalogue, Lacaille divided up the unwieldy constellation Argo Navis, the ship, into the subsections Carina, Puppis, and Vela that astronomers still use as separate constellations. As well as creating 14 new constellations, Lacaille eliminated a pre-existing one – Robur Carolinum, Charles's Oak, introduced by the Englishman Edmond Halley in 1678 to honour his monarch King Charles II.

All those from Lacaille's time onwards who gerrymandered the constellations did so without lasting success, but there were plenty of astronomers who tried to leave their mark on the sky. Constellation mania had reached its height by 1801 when the German astronomer Johann Elert Bode (1747–1826) published his immense star atlas, *Uranographia*, containing over 100 different constellations; but by then astronomers realized that things had gone too far, and during the ensuing century this number was eroded by a process of natural wastage. The English astronomer Francis Baily (1774–1844) was influential in whittling down the number; his *British Association Catalogue* of 1845 included 87 constellations, the only omission from the modern total being Hevelius's Scutum. In 1899 the American historian Richard Hinckley Allen summed up the prevailing situation in his book *Star Names and Their Meaning*: 'From 80 to 90 constellations may be considered as now more or less acknowledged'.

1.10. *The final 88*

The matter was settled once and for all by astronomy's newly founded governing body, the International Astronomical Union (IAU). One of the tasks undertaken by the IAU at its first General Assembly in 1922 was to agree a definitive list of 88 constellations covering the entire sky, with standardized names and three-letter abbreviations. The IAU did not explain how the final choice came about, but the names are the same as those in the leading star catalogue of the day, the *Revised Harvard Photometry*, published by Harvard College Observatory in 1908, so it seems that the IAU simply adopted those.

One serious deficiency remained, though: the constellations still had no properly defined boundaries. Since Bode's time cartographers had drawn free-hand lines snaking vaguely between constellation figures, but these were arbitrary and varied from atlas to atlas. What's more, some stars were shared between constellations, a tradition that extended back to the catalogue in Ptolemy's *Almagest*. Some form of standardization was needed. The Belgian astronomer Eugène Joseph Delporte (1882–1955) of the Royal Observatory in Brussels presented proposals for a clearly defined system of constellation boundaries to the IAU's second General Assembly in 1925, and the IAU gratefully handed him the task of turning them into reality.

Delporte drew his boundaries along lines of right ascension and declination for the year 1875. This date was chosen for consistency with the earlier work of the American astronomer Benjamin Apthorp Gould (1824–96), who in 1877 had published boundaries for the southern constellations in his atlas called *Uranometria Argentina*. Delporte's boundaries zig-zagged to ensure that all named variable stars remained within the constellations to which they were already assigned, as requested by the IAU's Variable Stars committee. Delporte also modified some of Gould's boundaries, particularly in places where he had used diagonal lines rather than verticals and horizontals.

Delporte's work, approved by the IAU at its meeting in 1928 and published in 1930 in a book called *Délimitation Scientifique des Constellations*, amounts to an international treaty on the demarcation of the sky, to which astronomers throughout the world have conformed ever since. Constellations are now regarded not as star patterns but as precisely defined areas of sky, rather like countries on Earth. Unlike the map of the Earth, though, the map of the sky is unlikely to change.

CHAPTER TWO

Star maps

EVERYONE is familiar with maps of the Earth, but to most people a map of the sky is a mystery. Yet there are many similarities because the celestial cartographer faces the same problem as the terrestrial one: how to represent a curved surface on a flat sheet.

The earliest representations of the sky were actually globes, on which the constellations were shown as though viewed from a God-like position beyond the stars; this meant that the constellation shapes were represented back to front by comparison with the way we see them from Earth. In the Museo Archeologico Nazionale, Naples, is a marble statue of Atlas holding on his shoulders a globe of the heavens on which the constellations are depicted in this mirror-image way (see Fig. 3). The sculpture is called the Farnese Atlas, after Cardinal Alessandro Farnese (later Pope Paul III) who acquired it in the early 16th century and exhibited it in the Farnese Palace in Rome.

This is the oldest known celestial globe, probably made in Rome in the second century AD. Even more significantly, the sculpture is thought to be a copy of a Greek original from the second or third century BC, which was around the time that Aratus wrote his *Phaenomena*. Thus the globe held by the Farnese Atlas provides our only firsthand look at the star pictures that the ancient Greeks imagined in the sky.

2.1. *Flat star charts*

An early form of flat star chart was the astrolabe, popular with the medieval Arabs. Usually made of brass, the astrolabe consisted of a flat base plate overlain by a rotating mask called the *rete*; pointers on the rete indicated the positions of prominent stars. Rotating the rete would show the positions of the stars at any given date or time; the same principle lives on in the star-finding devices called planispheres used by present-day amateur astronomers.

The earliest surviving astrolabes date from the tenth century AD but written evidence shows that they were known much earlier, possibly even in the time of Ptolemy around AD 150. Anywhere from a handful of stars to dozens could be included on the rete. In many cases, the pointers or their supports were engraved with the names of the stars, which could be written in Arabic or Latin depending where and by whom the astrolabe was made. Astrolabes are hence a rich source of old star names.

Other than astrolabes, the oldest known flat sky map is a Chinese paper scroll, over 2 metres long, thought to date from the mid to late 7th century AD.

Figure 3: The Farnese Atlas. This sculpture of Atlas from the second century AD shows him supporting a celestial globe engraved with the constellations known to the ancient Greeks. Unlike on later globes no individual stars are shown, just the constellation figures. On the left can be seen the stern of Argo Navis, the ship of the Argonauts (the prow is missing in the sky). Above and right of it is Hydra the water snake, with Crater and Corvus on its coils, while to the right is Centaurus. (Ilia Shurygin)

This scroll is known as the Dunhuang star chart after the place on the Silk Road trade route in north central China where it was found in the early 20th century; it is now in the British Library, London. Since the Dunhuang chart depicts the Chinese constellation tradition, which was independent of that in Europe and Arabia, most of the constellations are unrecognizable to modern eyes – see, for example, Fig. 4.

2.2. *Picturing the Ptolemaic constellations*

The earliest surviving depictions of the Ptolemaic constellations on paper are found in illuminated manuscripts of the poetic works of Aratus and Hyginus dating from the 9th century and onwards. These illustrations were the creation not of astronomers but of artists who interpreted the figures quite freely, with little concern for the framework of the underlying stars. As a result, the images bore only a loose resemblance to the constellation figures as described by Ptolemy. Woodcuts from later printed versions of these works, such as an edition of Hyginus's *Poetic Astronomy* published in Venice in 1482, are sometimes reproduced in modern books as examples of early constellation visualization, but they are not true star charts. Their intent was simply to provide a decorative accompaniment to literary texts, with no pretensions to scientific accuracy.

It was actually the Arabs who produced the first scientific depictions of the Ptolemaic constellations, over 800 years after the *Almagest* was written. Around the year 964 an Arabic astronomer named ʿAbd al-Raḥmān al-Ṣūfī (903–986), usually known simply as al-Ṣūfī (or Azophi in Latinized form), produced a revised and updated version of the star catalogue in the *Almagest* called the *Book of the Fixed Stars* (*Kitāb al-Kawākib al-Thābita* in Arabic). Al-Ṣūfī added charts of each constellation, something that the *Almagest* lacked (see, for example, Fig. 5). As well as accurately plotting the stars in the catalogue al-Ṣūfī provided two drawings of each constellation figure, one as it is seen in the sky and one reversed right to left as it would appear on a celestial globe. However, the Arabs adapted the imagery to their own culture: the human characters were clad in Arabic robes, not Greek ones, and there are other aspects of the iconography, such as the design of the ship Argo, that look unfamiliar to western eyes. So to find the true genesis of present-day star charts we must look back to Europe.

Figure 4: Chinese constellations differed markedly from Western ones, being usually much smaller and incorporating fainter stars. This illustration of the sky around the north celestial pole comes from a paper scroll dating from between AD 649 and 684 that was found in the early 20th century in caves at Dunhuang (pronounced dunn-hwong), China. Among the Chinese constellations in this region only the familiar shape of the Plough or Big Dipper is familiar to western eyes (below centre). The Dunhuang manuscript is the oldest surviving star map in the world. (British Library. MS Stein 3326)

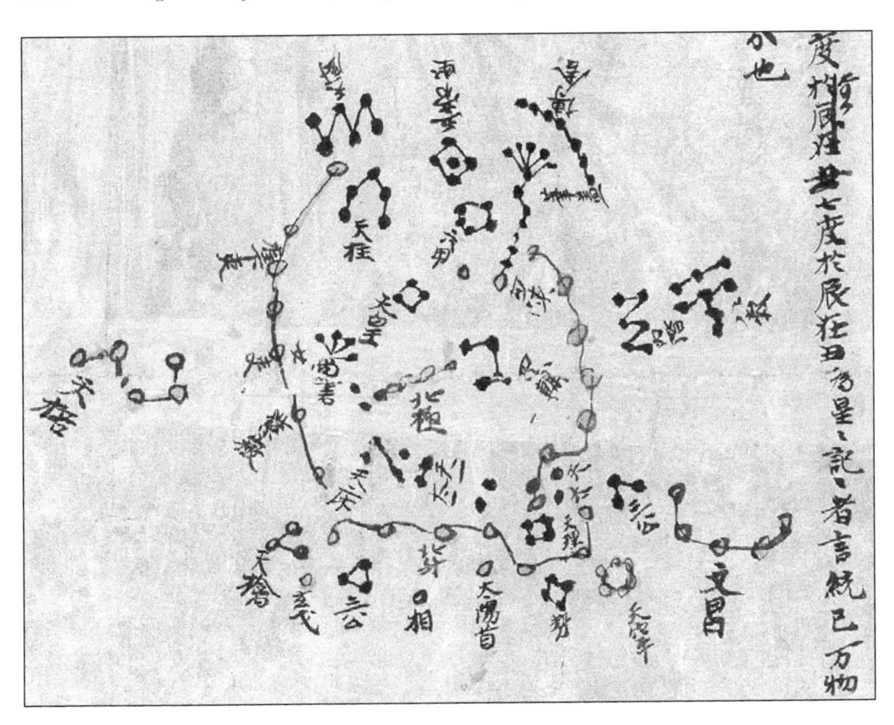

THE CHINESE SKY
A lost tradition

Chinese constellations were smaller than Western ones, and more numerous. Each constellation usually consisted of only a handful of stars which made it easier to specify areas of sky without the need for accurate coordinates. By the end of the third century AD, Chinese astronomers had developed an elaborate system of 283 constellations consisting of a total of 1464 stars.

These constellations did not depict myths but facets of Chinese imperial, social, and rural life. For example, we find *Dizuo*, the throne of the emperor (the star Alpha Herculis); *Huanzhe*, the court eunuchs (60 Herculis and three neighbouring stars); *Lingtai*, the astronomical observatory; and even the toilet, *Ce*, behind a modesty screen, *Ping*, both in Lepus. In some parts of the sky, groups of constellations on a common theme formed large tableaux depicting scenes such as the autumn harvest, the winter hunt, a cavalry camp, and a celestial market. Some of the same characters popped up repeatedly in different parts of the sky, notably the Emperor and members of his extensive retinue.

Unlike the imaginative artistry of western celestial cartography, Chinese star charts did not offer pictorial representations of the constellations. Instead, the chartmakers simply plotted the stars as dots of similar sizes connected by lines, with no attempt to scale the symbols according to the stars' brightnesses (see Fig. 4). This lack of a magnitude scale on Chinese charts adds to the difficulty of identifying the stars involved. The Chinese constellation system was unknown in the West and had no influence on the 88 celestial figures that we know today. It was still in use when Jesuit missionaries introduced Western constellations to China in the 17th century, after which it died out.

2.3. *First printed star chart*

A major step towards standardization of imagery of the Greek constellations came in 1515 when Albrecht Dürer (1471–1528), the great German artist, drew the first European printed chart of the heavens with the help of Johannes Stabius (*c*.1460–1522), an Austrian mathematician, and Conrad Heinfogel (d.1517), a German astronomer. Prior to this, all European and Arab star charts were individually hand-drawn, and hence restricted to single copies. With the advent of printing, large numbers of identical copies could be run off at will.

The chart consisted of a pair of woodcuts, one showing the zodiac and all constellations north of it, the other showing all known constellations south of the zodiac (see Fig. 1, pages 10–11). Both halves are based on the stars and constellations catalogued by Ptolemy in his *Almagest*, with positions updated by Heinfogel for the year 1500. The constellations are in mirror image, as on a celestial globe or an astrolabe, a tradition that most early maps were to follow. There are not many constellations in the southern half of the chart, for the far southern skies had not then been charted by Europeans. In the corners of the

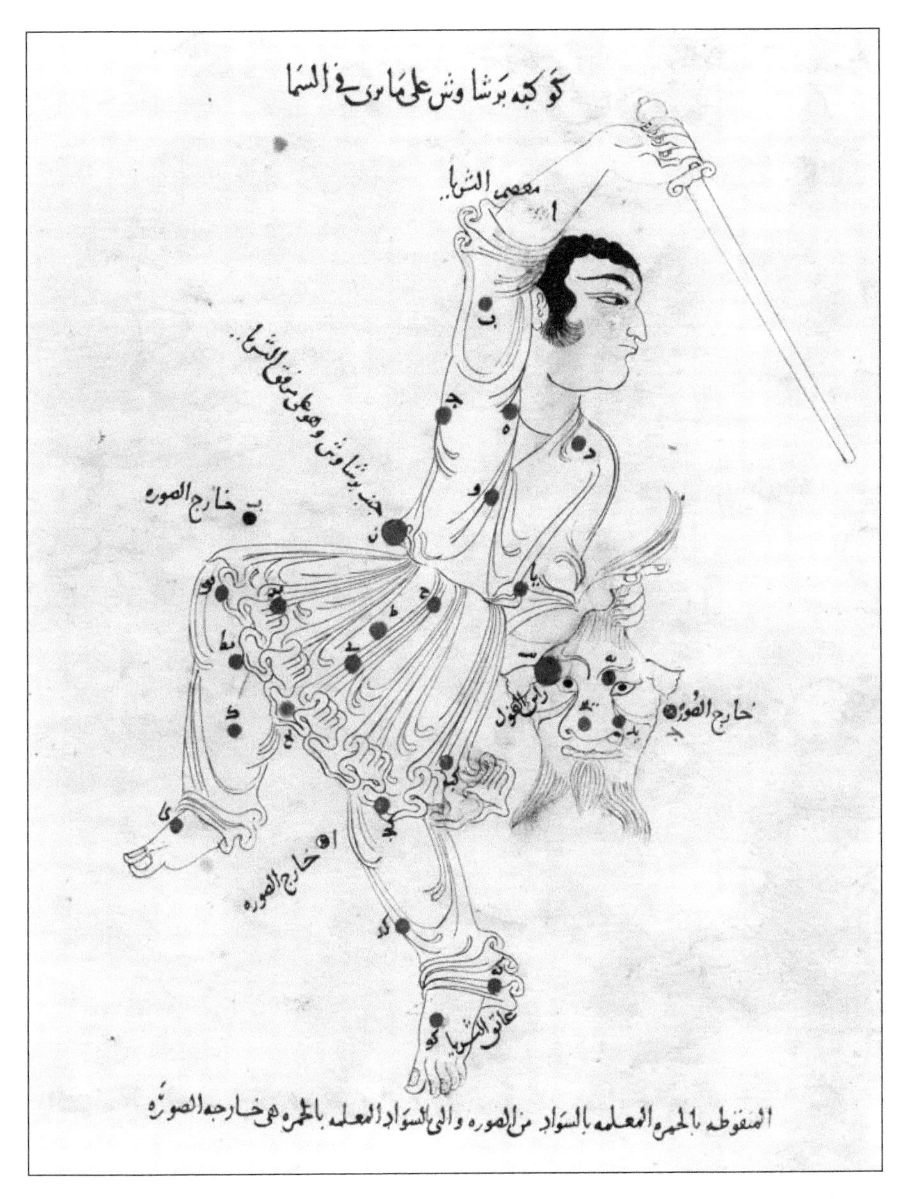

Figure 5: An Arabic representation of Perseus, from a version of the *Book of the Fixed Stars* by the Arabic astronomer al-Ṣūfī, also known by the Latinized name Azophi. This particular manuscript was reputedly made by his son around AD 1010, not long after al-Ṣūfī's death. The manuscript contains two illustrations of each constellation, one showing it as it appears in the sky and the other in reverse, as it would appear on a star globe; this plate shows Perseus as he appears in the sky. Perseus wears Arabic dress, and the head of Medusa the Gorgon has been depicted as a bearded male. The dotted object in the sword arm of Perseus, just above his sleeve, marks a twin cluster of stars known to modern astronomers as the Double Cluster, recorded by Ptolemy in the *Almagest* as a 'nebulous mass'. (Bodleian Library, Oxford. MS Marsh 144)

northern chart Dürer depicted the four ancient authorities on whose descriptions the constellation figures are based. Clockwise from top left they are Aratus, Ptolemy, al-Ṣūfī, and Manilius. Dürer's depictions of the constellation figures established an artistic style that was echoed on many later celestial maps. It, too, was superseded by a sequence of four great star atlases that in turn set new standards, both scientific and artistic.

2.4. *Bayer's Uranometria*

Star maps improved as astronomers surveyed the sky in more detail and with greater accuracy. In 1603 Johann Bayer (1572–1625), a lawyer in Augsburg, Germany, who had a passion for astronomy, published the first major printed star atlas, *Uranometria* (or, to give it its full title, *Uranometria, omnium asterismorum continens schemata, nova methodo delineata, aereis laminis expressa*, meaning 'Uranometria, containing charts of all the constellations, drawn by a new method and engraved on copper plates').

Uranometria devoted one large chart to each of the 48 Ptolemaic constellations. Bayer's main source of star positions and brightnesses was the recently

Figure 6: Johann Bayer's landmark star atlas of 1603 called *Uranometria* devoted individual charts to each of the 48 Greek constellations. Here Hercules is seen holding a branch from the golden apple tree of the Hesperides. Bayer's *Uranometria* was highly popular on account of its comprehensiveness, its artistic quality, and its introduction of the system of labelling stars with Greek letters. (David Rumsey Map Collection)

released catalogue by the great Danish observer Tycho Brahe (1546–1601), supplemented by the *Almagest* and observations of his own. The southern skies were allocated one map, depicting the 12 new constellations that had been invented a few years earlier from observations by the Dutch navigator Pieter Dirkszoon Keyser (see Section 1.7, page 19). This was the first appearance of these southern constellations in an atlas. In all, over 2,000 stars were plotted in the *Uranometria*, twice as many as shown by Dürer. So popular was Bayer's atlas that it was reprinted eight times between 1624 and 1689; its exquisitely engraved charts by the German painter and engraver Alexander Mair (*c.*1559–*c.*1620) of Augsburg are true works of art.

Bayer's atlas was notable for another reason: it introduced the convention of labelling bright stars by Greek letters, a system that astronomers still use. They are now commonly termed Bayer letters. For example, on this scheme the bright star Betelgeuse is also known as Alpha (α) Orionis, meaning Alpha of Orion (the genitive, or possessive, case of the constellation name is always used). Contrary to popular belief Bayer did not letter the stars in strict order of brightness – in fact, magnitude estimates at that time were not good enough for this to have been possible. What he actually did was to group the stars into magnitude classes, from first to sixth, then allocated letters to the members of each class as he saw fit.

For example, in Ursa Major the seven stars of the Plough were labelled in order of right ascension (the equivalent of longitude in the sky). In Gemini, the three brightest stars were labelled by declination, from north to south; while in Cygnus the sequence of letters for the brightest stars follows the overall shape of the constellation figure. In many other constellations, particularly among the fainter stars, there is no obvious pattern to the distribution of letters at all. As a result of this somewhat haphazard process there are actually 16 constellations in Bayer's atlas in which the star labelled Alpha is not the brightest: Cancer, Capricornus, Cetus, Corvus, Crater, Delphinus, Draco, Gemini, Hercules, Libra, Orion, Pegasus, Pisces, Sagitta, Sagittarius, and Triangulum.

Bayer did not assign Greek letters to the southern constellations of Keyser, perhaps reasoning that such a move would be premature. The Bayer lettering system was extended to the southernmost sky 160 years later by the French astronomer Nicolas Louis de Lacaille on his map published in 1763. Constellations in the northern sky that were introduced subsequent to Bayer's time were allocated Greek letters by the English astronomer Francis Baily (of Baily's beads fame) in his *British Association Catalogue* of 1845.

2.5. *Hevelius and the Firmamentum Sobiescianum*

A few years after Bayer's *Uranometria* appeared astronomy was revolutionized by the invention of the telescope, which not only showed faint stars that had hitherto been invisible but also greatly improved the accuracy with which star positions could be measured. One man remained unmoved by this advance: Johannes Hevelius, an astronomer from Danzig (the modern Gdańsk in Poland). Stubbornly, Hevelius continued to measure star positions with naked-eye sights throughout his life, worrying that lenses might introduce positional distortions.

Figure 7: Johannes Hevelius's influential star atlas *Firmamentum Sobiescianum* was published posthumously in 1690. Hevelius introduced ten new constellations, of which seven are still accepted by astronomers today. His atlas portrayed the constellation figures from the rear, as they would appear on a celestial globe. This led to some awkward-looking poses, such as Auriga carrying the goat and kids on his back, shown here. (ETH-Bibliothek Zürich)

Hevelius's catalogue of over 1,500 star positions, *Catalogus Stellarum Fixarum*, was published posthumously in 1690 as part of a book called *Prodromus Astronomiae*. The catalogue contained 50% more stars than that compiled by the great Tycho Brahe a century earlier and the positional measurements were of comparable accuracy. Accompanying Hevelius's catalogue was an atlas, *Firmamentum Sobiescianum*, engraved with great skill by the French engraver Charles de la Haye (1641–17??).

In this atlas and catalogue Hevelius introduced ten new constellations in the northern sky. For the southern stars Hevelius used the observations made by the English astronomer Edmond Halley (1656–1742) from the island of St Helena in the south Atlantic, which were an improvement on those of the pioneering Dutchmen Pieter Dirkszoon Keyser and Frederick de Houtman.

Firmamentum Sobiescianum suffers from the drawback that the constellation figures are depicted back to front, as they would appear on a celestial globe – see Fig. 7 for an example; this makes it difficult for an observer to match up the star patterns to the real sky. For this reason the Hevelius maps are not used for the constellation illustrations in this book.

2.6. *Flamsteed's Atlas Coelestis*

Celestial mapping took another major stride in the 18th century with the work of the first Astronomer Royal of England, John Flamsteed (1646–1719), who catalogued nearly 3,000 stars with unprecedented precision from the newly founded Royal Observatory at Greenwich. Flamsteed's star catalogue was published posthumously in 1725 in Volume 3 of his *Historia Coelestis Britannica*; the catalogue section itself was called *Catalogus Britannicus*. Four years later came *Atlas Coelestis*, a set of 25 elegantly engraved celestial charts based entirely on Flamsteed's own observations. The far southern skies, below the horizon of Greenwich, are covered by one small chart based on the observations made by Halley at St Helena. This southern chart depicts the 12 constellations of Keyser and de Houtman plus Halley's own invention, Robur Carolinum, now obsolete.

Flamsteed's stars were divided into 55 constellations. Two Ptolemaic constellations, Ara and Corona Australis, were omitted because they were too far south for him to see. He accepted six of the constellations invented by Hevelius (Canes Venatici, Lacerta, Leo Minor, Lynx, Sextans, and Vulpecula), whereas today's sky includes a seventh, Scutum; this was reinstated by Johann Bode in his *Uranographia* of 1801. Flamsteed's remaining three non-Ptolemaic constellations were Camelopardalis, Coma Berenices, and Monoceros.

In the *Atlas Coelestis*, Flamsteed took particular care to depict the Greek constellation figures exactly as Ptolemy had described them. The introduction to the atlas contains some disapproving words about the way that Bayer had represented the constellation figures in his *Uranometria*:

> Having drawn all his human figures, except Boötes, Andromeda and Virgo, with their backs towards us, those stars, which all before him place in the right shoulders, sides, hands, legs or feet, fall in the left, and the contrary ... whereby he renders the oldest observations false or nonsense.

Despite popular misconception, Flamsteed did not introduce the so-called Flamsteed number system for identifying the stars in each constellation; that was done in 1783 by a French astronomer, Joseph Jérôme de Lalande (1732–1807). In a French edition of Flamsteed's catalogue, published in an almanac called *Éphémérides des mouvemens célestes*, Lalande inserted a column in which he numbered the stars consecutively in each constellation in the order that Flamsteed had listed them, and this is the system that astronomers mean when they speak of Flamsteed numbers. Stars are usually referred to by their Flamsteed numbers – for example 61 Cygni or 70 Ophiuchi – only when they are not already identified by a Greek letter.

One legacy of Flamsteed's atlas which is sometimes overlooked is the sequence of smaller popular atlases that it inspired: Jean Fortin's *Atlas Céleste* in France (1776 and 1795), Johann Bode's *Vorstellung der Gestirne* in Germany (1782 and 1805), and Alexander Jamieson's *Celestial Atlas* in England (1822), all of which in turn had their own imitators. Jamieson's atlas in particular was closely copied for a famous set of constellation cards called *Urania's Mirror* issued two years later. The constellation figures in Elijah Burritt's *Atlas of the Heavens* of 1835, a highly popular American star atlas that went through many subsequent editions, were also closely modelled on those of Jamieson.

2.7. *Bode's Uranographia*

Flamsteed's catalogue and atlas set new standards in astronomy, and I have used his atlas as one of the sources for illustrations on these pages. The other main source is the greatest of the old-style pictorial star atlases, *Uranographia*, published in 1801 by the German astronomer Johann Elert Bode, director of Berlin Observatory. (It actually appeared in five parts from 1797 onwards, but 1801 was the completion date.)

Bode's *Uranographia* was the first atlas to depict virtually all the stars visible to the naked eye (i.e. down to sixth magnitude), plus a fair selection of those down to eighth magnitude, six times fainter. Over 17,000 stars are plotted, taken from the observations of various astronomers including Flamsteed, Lacaille, Lalande, and Bode himself. To accompany the atlas Bode produced a catalogue called *Allgemeine Beschreibung und Nachweisung der Gestirne*, also published in 1801.

Uranographia was also the first major star atlas with boundary lines drawn between the constellations, albeit vague and ill-defined, unlike the rigorous modern boundaries. Even Flamsteed, normally so punctilious, had been entirely unspecific about the extent of each constellation on his charts.

Bode intended the *Uranographia* to be comprehensive and he certainly succeeded, for in addition to charting a greater number of stars than any previous cartographer he also depicted more constellations – over 100 in all. Among them were five constellations making their debut on this atlas: Felis and Globus Aerostaticus were both suggested by Lalande during the preparation of the atlas, while Lochium Funis, Machina Electrica, and Officina Typographica were invented by Bode himself. None of these five survived the test of time, however. For more about them see Chapter Four.

Bode's *Uranographia* marked the end of an era. Thereafter, astronomers placed decreasing emphasis on the fanciful (and physically meaningless) constellation figures of the Greeks, concentrating instead on the exact measurement of position, brightness, and physical properties of the stars.

2.8. *End of a tradition*

In the transition from classical to scientific mapping that occurred during the 19th century one atlas stands out: the *Uranometria Nova* of the German astronomer Friedrich Wilhelm August Argelander (1799–1875) published in 1843. In this, the constellation figures, printed in red, were reduced to shadowy insignificance by comparison with the stars. This same two-colour style was followed by Argelander's countryman Eduard Heis (1806–77) in his *Atlas Coelestis Novus* of 1872. These atlases were the standard references for professional astronomers of the day, and their choice of constellations helped establish the eventual list of 88 adopted by the International Astronomical Union in 1922.

By the end of the 19th century, two thousand years of Greek tradition had finally given way to the facts-and-figures approach of astronomical census-takers and statisticians. Where the ancient Greeks imagined their gods and heroes populating the sky, modern astronomers have discovered the existence of an equally fantastic pantheon of objects with names such as red giants, white dwarfs, Cepheid variables, pulsars, quasars, and black holes.

CHAPTER THREE

The celestial eighty-eight

BIOGRAPHIES of all 88 constellations officially sanctioned by the International Astronomical Union are found on the following pages, plus one for the Milky Way. Each entry describes the mythology and history of the constellation along with an illustration of it from a classic star atlas, usually Johann Bode's *Uranographia* of 1801 or John Flamsteed's *Atlas Coelestis of* 1729. For the more prominent constellations I have included some brief information about the brightest stars and other objects of interest they contain, including the origin and meaning of certain star names. For a selection of obsolete constellations, see Chapter Four.

Andromeda

Genitive: Andromedae
Abbreviation: And
Size ranking: 19th
Origin: One of the 48 Greek constellations listed by Ptolemy in the *Almagest*
Greek name: Ἀνδρομέδα (Andromeda)

Perhaps the most enduring of all Greek myths is the story of Perseus and Andromeda, the original version of George and the dragon. Its heroine is the beautiful princess Andromeda (Ἀνδρομέδα in Greek). She was the daughter of the weak King Cepheus of Ethiopia and the vain Queen Cassiopeia, whose boastfulness knew no bounds.

Andromeda's misfortunes began one day when her mother claimed that she was more beautiful even than the Nereids, a particularly alluring group of sea nymphs. The affronted Nereids decided that Cassiopeia's vanity had finally gone too far and they asked Poseidon, the sea god, to teach her a lesson. In retribution, Poseidon sent a terrible monster (some say also a flood) to ravage the coast of King Cepheus's territory. Dismayed at the destruction, and with his subjects clamouring for action, the beleaguered Cepheus appealed to the Oracle of Ammon for a solution. He was told that he must sacrifice his virgin daughter to appease the monster.

Hence the blameless Andromeda came to be chained to a rock to atone for the sins of her mother, who watched from the shore with bitter remorse. The

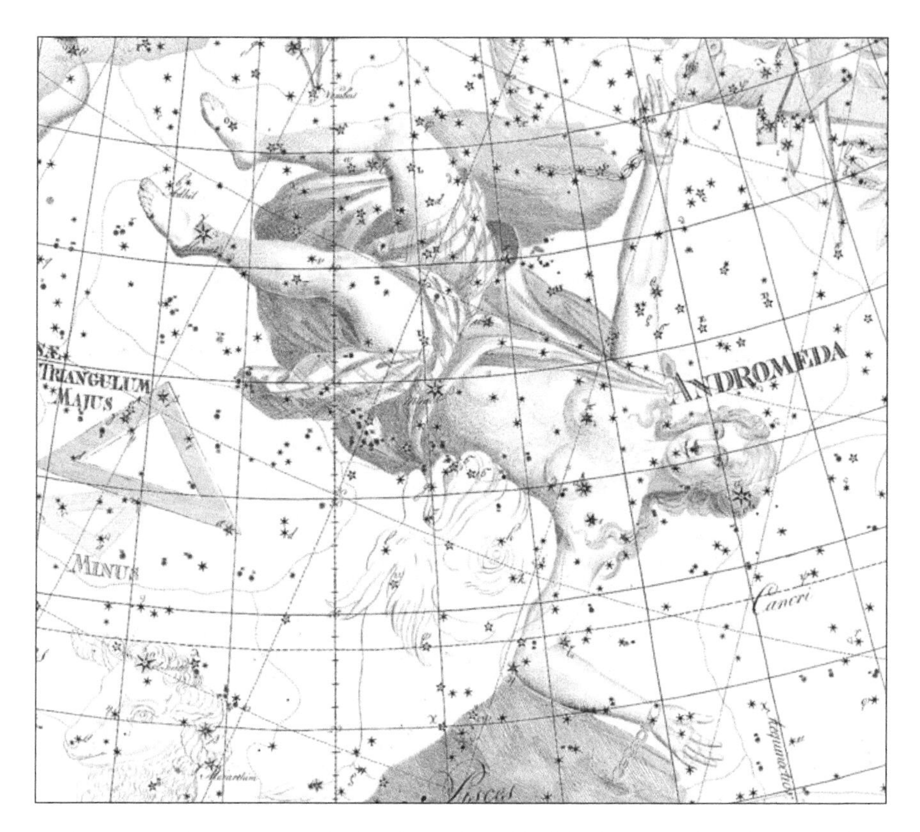

Andromeda chained to a rock, depicted on Chart IV of the *Uranographia* of Johann Bode (1801). (Deutsches Museum)

site of this event is said to have been on the Mediterranean coast at Joppa (Jaffa), the modern Tel-Aviv. As Andromeda stood on the wave-lashed cliffs, pale with terror and weeping pitifully at her impending fate, the hero Perseus happened by, fresh from his exploit of beheading Medusa the Gorgon. His heart was captivated by the sight of the frail beauty in distress below.

The Roman poet Ovid tells us in his book the *Metamorphoses* that Perseus at first almost mistook her for a marble statue. Only the wind ruffling her hair and the warm tears on her cheeks showed that she was human. Perseus asked her name and why she was chained there. Shy Andromeda, totally different in character from her vainglorious mother, did not at first reply; even though awaiting a horrible death in the monster's slavering jaws, she would have hidden her face modestly in her hands, had they not been bound to the rock.

Perseus persisted in his questioning. Eventually, afraid that her silence might be misinterpreted as guilt, she told Perseus her story, but broke off with a scream as she saw the monster breasting through the waves towards her. Pausing politely to ask the permission of her parents for Andromeda's hand in marriage, Perseus swooped down, slew the sea-dragon with his diamond sword, released the swooning girl to the enthusiastic applause of the onlookers, and claimed her for his bride. Andromeda later bore Perseus six children including Perses, ancestor of the Persians, and Gorgophonte, father of Tyndareus, king of Sparta.

The mythologists said that the Greek goddess Athene placed Andromeda's image among the stars, where she lies between Perseus and her mother Cassiopeia. Only the constellation Pisces, the fishes, separates her from the sea monster, Cetus.

Stars in Andromeda – and a spiral galaxy

Star maps picture Andromeda with her hands in chains. Her head is marked by the second-magnitude star Alpha Andromedae, originally shared with neighbouring Pegasus. Echoes of its former dual identity live on in its popular name of Alpheratz, which comes from the Arabic *al-faras*, meaning 'the horse'. Her waist is marked by the star Beta Andromedae, called Mirach, a name corrupted from the Arabic *al-mi'zar* meaning 'the girdle' or 'loin cloth'. Her left foot is marked by Gamma Andromedae, whose official IAU-approved name is Almach, but which in the past has also been variously spelled Almaak, Alamak, or Almak, from the Arabic *al-'anaq*, referring to the desert lynx or caracal which the old Arabs visualized here. Through small telescopes this is a beautiful twin star of contrasting yellow and blue colours.

The most celebrated object in the constellation is the great spiral galaxy M31, positioned on Andromeda's right hip, where it is visible as an elongated blur to the naked eye on clear nights. M31 is a whirlpool of stars similar to our own Milky Way. At a distance of 2.5 million light years, the Andromeda Galaxy is the farthest object visible to the naked eye. Discovery of this object is attributed to the Arabic astronomer al-Ṣūfī (903–986), or Azophi in Latinized form, who first mentioned it in his *Book of the Fixed Stars* (*c*. AD 964).

Antlia

The air pump

Genitive: Antliae
Abbreviation: Ant
Size ranking: 62nd
Origin: The 14 southern constellations of Nicolas Louis de Lacaille

Antlia is one of the constellations of the southern sky introduced by the French astronomer Nicolas Louis de Lacaille on his chart of 1756. He said that it symbolized experimental physics. Lacaille originally called the constellation la Machine Pneumatique but Latinized the name to Antlia Pneumatica on the second edition of his map published in 1763. Following a suggestion by John Herschel (1792–1871), the English astronomer Francis Baily shortened its name to just Antlia in his *British Association Catalogue* of 1845, and it has been known as that ever since. Lacaille depicted it as the single-cylinder type of pump used by the French physicist Denis Papin during the early 1670s for his experiments

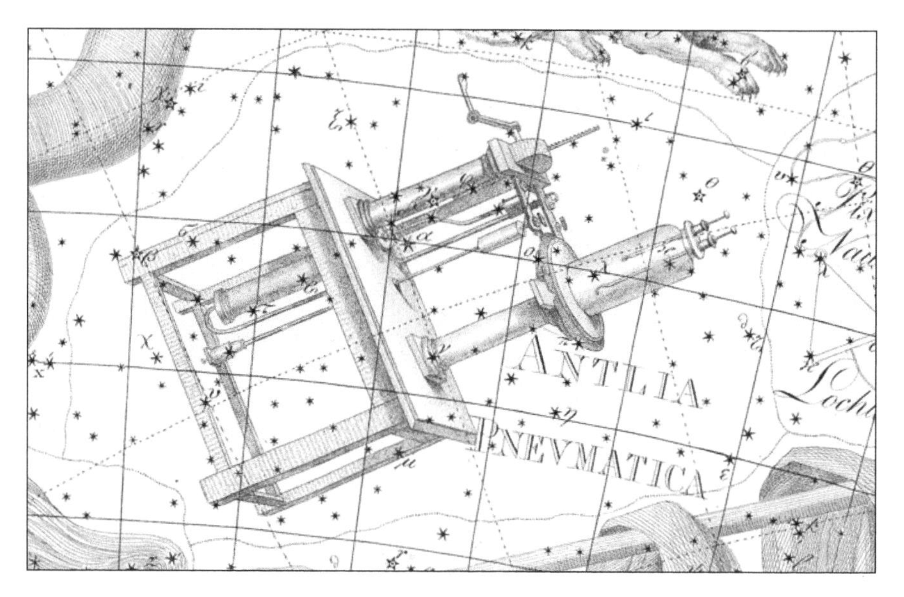

The air pump shown as a complex piece of apparatus on Chart XIX of the *Uranographia* of Johann Bode (1801). Air pumps became scientific toys for the rich during the 18th century. Here, the mast of Argo Navis is seen to the south of Antlia, with the feet of Felis the cat, a now-obsolete Bode invention, visible at the top. (Deutsches Museum)

on vacuums. In 1675 Papin moved from Paris to London where he worked with the Irish physicist Robert Boyle. Here Papin developed the more efficient double-cylinder type of pump, and it is one of these later types of pump that was illustrated by Johann Bode in his *Uranographia* star atlas of 1801. An air pump of this type is seen in action in the famous painting titled *An Experiment on a Bird in the Air Pump* by Joseph Wright of Derby (1768).

There are no legends associated with Antlia. Its brightest star, Alpha Antliae, is only of fourth magnitude and it contains no objects of note. Its name, however, is one to catch the unwary as it is frequently mis-spelled 'Antila'.

Apus

The bird of paradise

Genitive: Apodis
Abbreviation: Aps
Size ranking: 67th
Origin: The 12 southern constellations of Keyser and de Houtman

One of the dozen new constellations introduced at the end of the 16th century from observations of the southern sky by the Dutch navigators Pieter Dirkszoon Keyser and Frederick de Houtman. Apus represents a fabulous bird of paradise, as found in New Guinea, but it is a disappointing tribute to such an exotic creature, its brightest stars being of only 4th magnitude.

The name Apus comes from the Greek *apous*, meaning 'footless', since the birds were originally known to westerners only from dead specimens without feet or wings; these appendages had been removed by the locals, who prized the plumage for ornamental dress. The first examples were brought back to Europe by the survivors of Ferdinand Magellan's round-the-world voyage in 1522, creating immense interest. For a while it was speculated that these gaudy birds were the mythical phoenix.

The constellation Apus was first shown on the 1598 celestial globe of Petrus Plancius as 'Paradysvogel Apis Indica'. It seems likely that the word 'apis', meaning bee, was a misprint for 'avis', meaning bird, particularly since in that same year the Dutchman Jan van Linschoten had given the birds the Latin name *Avis paradiseus*. Johann Bayer also called the constellation Apis Indica on

Apus seen on Chart XX of the *Uranographia* of Johann Bode (1801), where it was given the alternative title of Avis Indica, the Indian bird, referring to its habitat of the East Indies. The bird's tail originally extended closer to the south celestial pole at lower left, but was clipped by Lacaille in the 1750s to make room for Octans. (Deutsches Museum)

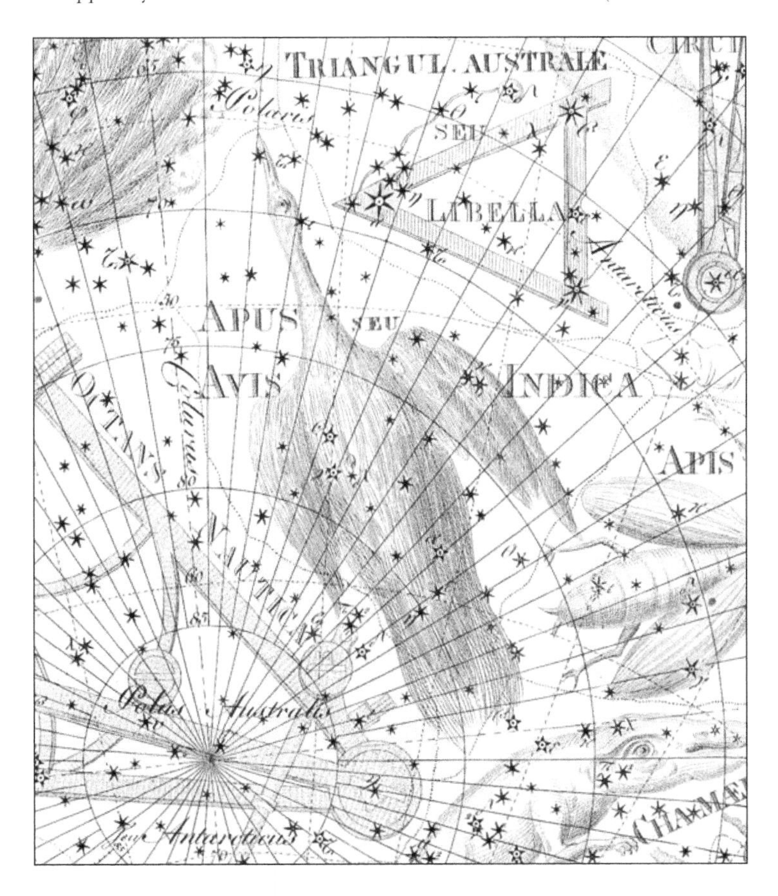

his *Uranometria* atlas of 1603, where it is depicted without wings or feet, doubtless modelled on a dead specimen. Others, such as Johannes Kepler in the *Rudolphine Tables* of 1627, called it 'Apus, Avis Indica' (Apus, bird of India), correcting the apparent misprint, but the alternative usages of Apis and Avis continued to Bode's day.

Part of the bird's tail was docked by Lacaille in the 1750s to form his south polar constellation Octans, an unfortunate truncation given that in real life the long, colourful tail feathers are the bird's main attraction. Apus has no named stars, nor are there any legends associated with it.

Aquarius

The water carrier

Genitive: Aquarii
Abbreviation: Aqr
Size ranking: 10th
Origin: One of the 48 Greek constellations listed by Ptolemy in the *Almagest*
Greek name: Ὑδροχόος (Hydrochoös)

Star maps show Aquarius as a young man pouring water from a jar or amphora, although Ovid, in his *Fasti*, says the liquid is a mixture of water and nectar, the drink of the gods. The water jar is marked by a Y-shaped asterism of four stars centred on Zeta Aquarii, and the stream ends in the mouth of the southern fish, Piscis Austrinus. Who is this young man commemorated as Aquarius?

The most popular identification is that he is Ganymede or Ganymedes, said to have been the most beautiful boy alive. He was the son of King Tros, who gave Troy its name. One day, while Ganymede was watching over his father's sheep, Zeus became infatuated with the shepherd boy and swooped down on the Trojan plain in the form of an eagle, carrying Ganymede up to Olympus (or, according to an alternative version, sent an eagle to do it for him). The eagle is commemorated in the neighbouring constellation of Aquila. As compensation for spiriting away his son, Zeus presented King Tros with a pair of fine horses, although some writers said the gift was a golden vine.

In another version of the myth, Ganymede was fought over by two rival admirers: he was first carried off by Eos, goddess of the dawn, who had a passion for young men, but was then stolen from her by omnipotent Zeus. Either way, Ganymede became wine-waiter to the gods of Olympus, dispensing nectar from his bowl, to the annoyance of Zeus's wife Hera. Robert Graves tells us that this myth became highly popular in ancient Greece and Rome where it was regarded as signifying divine endorsement for homosexuality. The Latin translation of the name Ganymede, Catamitus, gave rise to the word catamite.

If this myth seems insubstantial to us, it is perhaps a result of the Greeks imposing their own story on a constellation they adopted from elsewhere. The constellation of the water pourer originally seems to have represented the

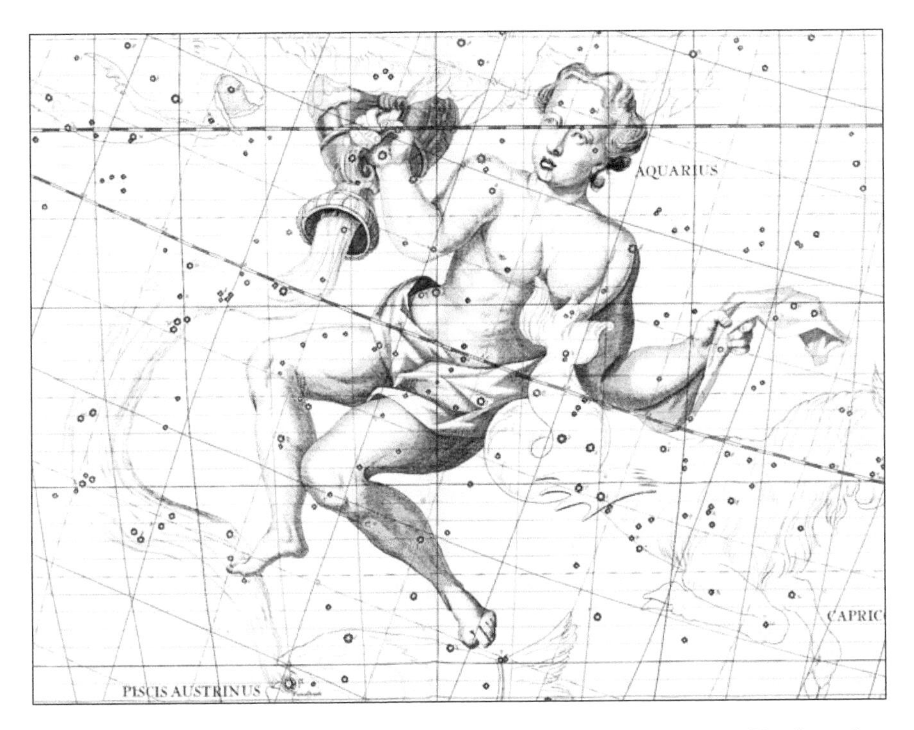

Aquarius and his water jar, from the *Atlas Coelestis* of John Flamsteed (1729). The flow of water from the jar extends into the mouth of the southern fish, Piscis Austrinus, at the bottom of the chart. (University of Michigan Library)

Egyptian god of the Nile – but, as Robert Graves notes, the Greeks were not much interested in the Nile.

Germanicus Caesar identified the constellation with Deucalion, son of Prometheus, one of the few men to escape the great flood. 'Deucalion pours forth water, that hostile element he once fled, and in so doing draws attention to his small pitcher', wrote Germanicus. Hyginus offered an additional identification of the constellation: he said it was Cecrops, an early king of Athens, seen making sacrifices to the gods using water, for he ruled in the days before wine was made.

Stars of Aquarius

The constellation's name in Greek was Ὑδροχόος, i.e. Hydrochoös. In the *Almagest*, Ptolemy described Alpha, Beta, and Gamma Aquarii as lying in the right shoulder, left shoulder, and the right forearm of the figure respectively. The head of Aquarius is marked by the relatively lowly 25 Aquarii, of 5th magnitude. Ptolemy listed 20 stars as lying in the flow of water from the jar, almost as many stars as in the main figure of Aquarius itself. In fact, early writers such as Aratus and Eratosthenes regarded the Water (Hydor, Ὕδωρ) as either a separate constellation or a sub-constellation within Aquarius. The flow of water started at the star we know as Kappa Aquarii and ended at the star we call Fomalhaut, in the mouth of the southern fish, Piscis Austrinus. This is an example of stars being shared between constellations as happened in the days before rigorous boundaries were established. Fomalhaut is now the exclusive property of Piscis Austrinus.

Several stars in Aquarius have names of Arabic origin beginning with 'Sad'. In Arabic, *sa'd* means 'luck'. Alpha Aquarii is called Sadalmelik, from *sa'd al-malik*, usually translated as 'the lucky stars of the king'. Beta Aquarii is called Sadalsuud, from *sa'd al-su'ud*, possibly meaning 'luckiest of the lucky'. Gamma Aquarii is Sadachbia, from *sa'd al-akhbiya*, possibly meaning 'lucky stars of the tents'. The exact significance of these names has been lost even by the Arabs, according to the German expert on star names, Paul Kunitzsch.

Aquila
The eagle

Genitive: Aquilae
Abbreviation: Aql
Size ranking: 22nd
Origin: One of the 48 Greek constellations listed by Ptolemy in the *Almagest*
Greek name: Ἀετός (Aetos)

Aquila represents an eagle, the thunderbird of the Greeks. The constellation's name in Greek was Ἀετός, i.e. Aetos, meaning eagle. There are several explanations for the presence of this eagle in the sky. In Greek and Roman mythology, the eagle was the bird of Zeus, carrying (and retrieving) the thunderbolts which the wrathful god hurled at his enemies. But the eagle was involved in love as well as war.

According to one story, Aquila is the eagle that snatched up the beautiful Trojan boy Ganymede, son of King Tros, to become the cup-bearer of the gods on Olympus. Authorities such as the Roman poet Ovid say that Zeus turned himself into an eagle to effect the abduction, whereas others say that the eagle was simply sent by Zeus. Ganymede himself is represented by the neighbouring constellation of Aquarius, and star charts show Aquila swooping down towards Aquarius. Germanicus Caesar says that the eagle is guarding the arrow of Eros (neighbouring Sagitta) which made Zeus lovestruck.

The constellations of the eagle and the swan are linked in an account by Hyginus. Zeus fell in love with the goddess Nemesis but, when she resisted his advances, he turned himself into a swan and had Aphrodite pretend to pursue him in the form of an eagle. Nemesis gave refuge to the escaping swan, only to find herself in the embrace of Zeus. To commemorate this successful trick, Zeus placed the images of swan and eagle in the sky as the constellations Cygnus and Aquila.

The name of the constellation's brightest star, Altair, comes from the Arabic *al-nasr al-ta'ir*, meaning 'flying eagle' or 'vulture'. The Greeks, including Aratus

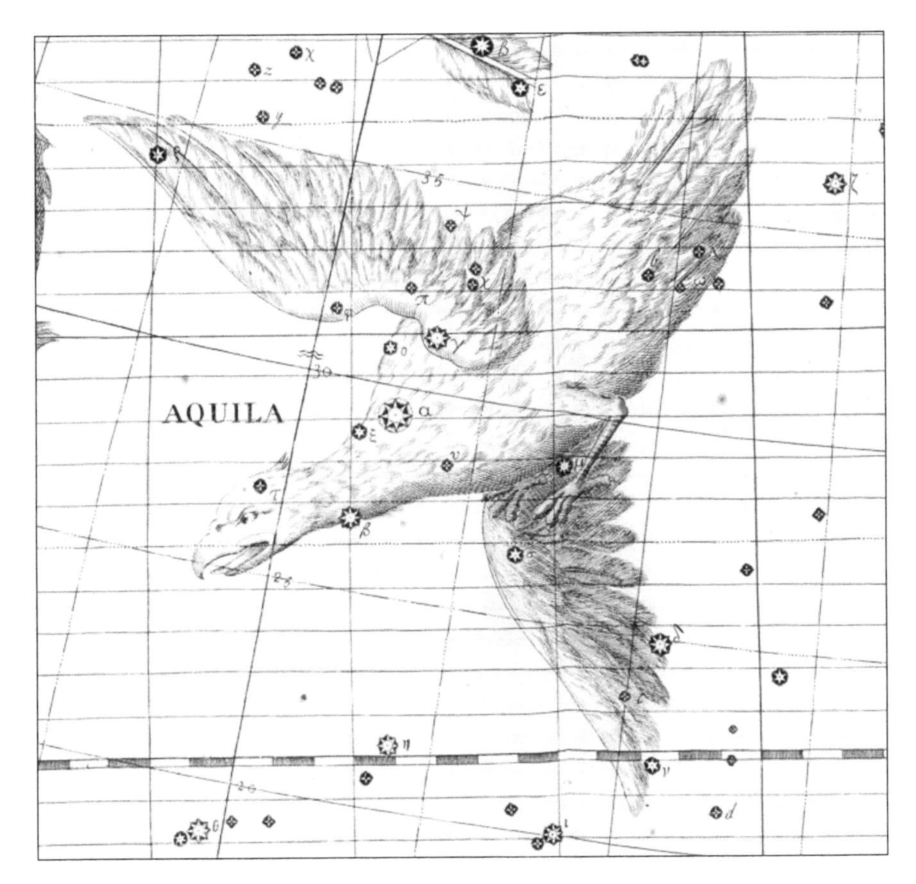

Aquila swoops across this chart from John Flamsteed's *Atlas Coelestis* (1729). Its brightest star, Altair, labelled Alpha, lies at the base of its neck. (University of Michigan Library)

and Ptolemy, called this star Ἀετός (Aetos), the eagle, the same as the whole constellation. The German scholar Paul Kunitzsch notes that the Babylonians and Sumerians referred to Altair as the eagle star, testimony to an even more ancient origin of the name. Altair's neighbouring stars Beta and Gamma Aquilae lie in the eagle's neck and in its left shoulder respectively, according to Ptolemy. These two stars have their own names, Alshain and Tarazed, which come from a Persian translation of an old Arabic word meaning 'the balance'.

Altair forms one corner of the so-called Summer Triangle with the stars Vega and Deneb, found in the constellations Lyra and Cygnus respectively. A charming Chinese folk tale visualizes Altair as a cowherd flanked by his two sons, separated by the Milky Way from his wife the weaving girl (the star Vega). The two were able to meet on just one day each year when magpies collect to form a bridge across the celestial river.

The southern part of Aquila was subdivided by Ptolemy into a now-obsolete constellation called Antinous, visualized on some maps as a young man being held in the eagle's claws (see page 186).

Ara

The altar

Genitive: Arae
Abbreviation: Ara
Size ranking: 63rd
Origin: One of the 48 Greek constellations listed by Ptolemy in the *Almagest*
Greek name: θυμιατήριον (Thymiaterion)

Altars feature frequently in Greek legend, for heroes were always making sacrifices to the gods, so it is not surprising to find an altar among the stars. But this altar is a special one, for it was used by the gods themselves to swear a vow of allegiance before their fight against the Titans, according to Eratosthenes and Manilius. That clash, known as the Titanomachy, was one of the most significant events in Greek myth.

At that time the ruler of the Universe was Cronus, one of the 12 Titans. Cronus had overthrown his father, Uranus, but it was prophesied that he would in turn be deposed by one of his own sons. In a desperate attempt to forestall the prophesy, Cronus swallowed his children as they were born: Hestia, Demeter, Hera, Hades, and Poseidon, all ultimately destined to become gods and goddesses. Eventually his wife, Rhea, could not bear to see any more children swallowed. She smuggled the next child, Zeus, to the cave of Dicte in Crete and gave Cronus a stone to swallow instead, telling him it was the infant Zeus.

On Crete, Zeus grew up safely. When he reached maturity he returned to his father's palace and forced Cronus to vomit up the children he had swallowed, who emerged as fully grown gods and goddesses. Zeus and his brother gods then

Ara, the altar, depicted as an elegant censer with its flames rising southwards in the *Uranometria* of Johann Bayer (1603). (ETH-Bibliothek Zürich)

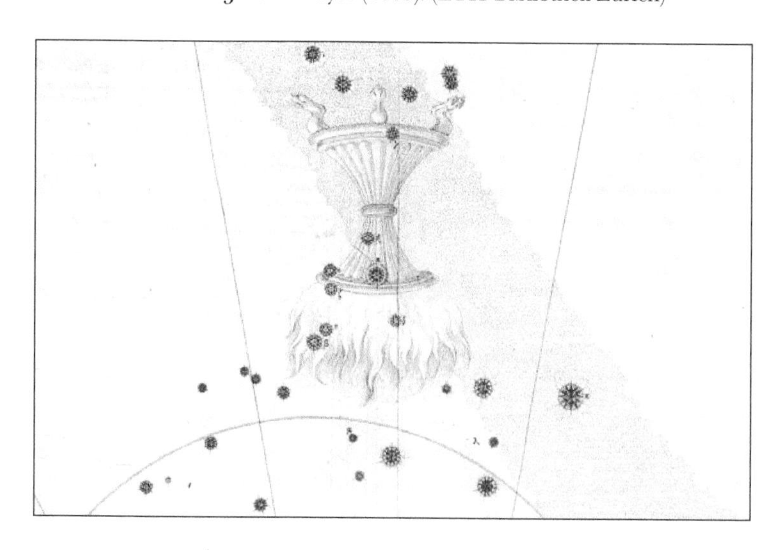

set up an altar and vowed on it to overthrow the callous rule of Cronus and the other Titans.

The battle raged for ten years between the Titans, led by Atlas, on Mount Othrys, and the gods led by Zeus on Mount Olympus. To break the deadlock Mother Earth (Gaia) instructed Zeus to release the ugly brothers of the Titans, whom Cronus had imprisoned in the sunless caves of Tartarus, the lowermost region of the Underworld. There were two sets of brothers, the Hecatoncheires (hundred-handed giants) and the one-eyed Cyclopes, and they wanted revenge against Cronus.

Zeus stole down to Tartarus, released the monstrous creatures and asked them to join him in the battle raging above. Delighted by their unexpected freedom, the Cyclopes set to work to help the gods. They fashioned a helmet of darkness for Hades, a trident for Poseidon and, above all, thunderbolts for Zeus. With these new weapons and their monstrous allies, the gods routed the Titans.

After their victory, the gods cast lots to divide up the Universe. Poseidon became lord of the sea, Hades won the Underworld and Zeus was allotted the sky. Zeus then placed the altar of the gods in the sky as the constellation Ara in lasting gratitude for their victory over the Titans.

The Greek name for the constellation was θυμιατήριον (Thymiaterion), referring to an incense burner or censer. Aratus shortened this to θυτήριον (Thyterion). An alternative Latin name widely encountered prior to the 18th century was Thuribulum, with the same meaning. The altar is usually depicted with its base to the north and its flames rising southwards, as described by Ptolemy in the *Almagest*. Some atlases depict Ara as the altar on which Centaurus is about to sacrifice Lupus, the wolf.

The Greeks regarded Ara as a sign of storms at sea. According to Aratus, if the altar was visible while other stars were covered by cloud, mariners could expect southerly gales.

Aries

The ram

Genitive: Arietis
Abbreviation: Ari
Size ranking: 39th
Origin: One of the 48 Greek constellations listed by Ptolemy in the *Almagest*
Greek name: Κριός (Krios)

It is not surprising to find a ram in the sky, for rams were frequently sacrificed to the gods, and Zeus was at times identified with a ram. But the mythographers agree that Aries is a special ram, the one whose golden fleece was the object of the voyage of Jason and the Argonauts. This ram made its appearance on Earth just as King Athamas of Boeotia was about to sacrifice his son Phrixus to ward off an impending famine.

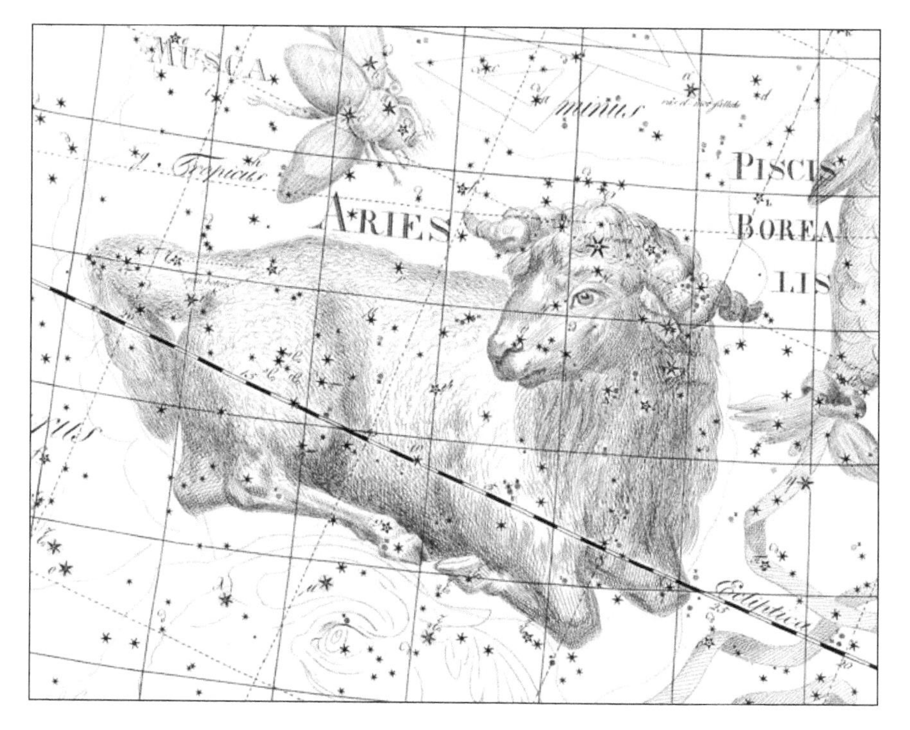

Aries, the ram with the golden fleece, from Chart XI of the *Uranographia* of Johann Bode (1801). In the sky, the ram is without its fleece, which was left behind on Earth. Above it is the now-obsolete constellation of Musca, the fly. (Deutsches Museum)

King Athamas and his wife Nephele had an unhappy marriage, so Athamas turned instead to Ino, daughter of King Cadmus from neighbouring Thebes. Ino resented her step-children, Phrixus and Helle, and she arranged a plot to have them killed. She began by parching the wheat so that the crops would fail. When Athamas appealed for help to the Delphic Oracle, Ino bribed messengers to bring back a false reply that Phrixus must be sacrificed to save the harvest.

Reluctantly, Athamas took his son to the top of Mount Laphystium, overlooking his palace at Orchomenus. He was about to sacrifice Phrixus to Zeus when Nephele intervened to save her son, sending down from the sky a winged ram with a golden fleece. Phrixus climbed on the ram's back and was joined by his sister Helle, who feared for her own life. They flew off eastwards to Colchis, which lay on the eastern shore of the Black Sea, under the Caucasus Mountains (the modern Georgia). On the way Helle's grip failed and she fell into the channel between Europe and Asia, the Dardanelles, which the Greeks named the Hellespont in her memory. On reaching Colchis, Phrixus sacrificed the ram in gratitude to Zeus. He presented its golden fleece to the fearsome King Aeëtes of Colchis who, in return, gave Phrixus the hand of his daughter Chalciope. Nephele placed the image of the ram among the stars.

Eratosthenes, though, said that the ram was immortal, which would have made sacrificing it highly problematic. Instead, in this version, the ram itself shed its fleece and flew into the sky of its own accord. Either way, the lack of the radiant fleece was said by the mythologists to account for the constellation's relative faintness.

After Phrixus died his ghost returned to Greece to haunt his cousin Pelias, who had seized the throne of Iolcus in Thessaly. The true successor to the throne was Jason. Pelias promised to give up the throne to Jason if he brought home the golden fleece from Colchis. This was the challenge that led to the epic voyage of Jason and the Argonauts.

When he reached Colchis, Jason first asked King Aeëtes politely for the fleece, which hung on an oak in a sacred wood, guarded by a huge unsleeping serpent. King Aeëtes rejected Jason's request. Fortunately for the expedition, the king's daughter, Medea, fell in love with Jason and offered to help him steal the fleece. At night the two crept into the wood where the golden fleece hung, shining like a cloud lit by the rising Sun. Medea bewitched the serpent so that it slept while Jason snatched the fleece.

According to Apollonius Rhodius, the fleece was as large as the hide of a young cow, and when Jason slung it over his shoulder it reached his feet. The ground shone from its glittering golden wool as Jason and Medea escaped with it. Once free of the pursuing forces of King Aeëtes, Jason and Medea used the fleece to cover their wedding bed. The final resting place of the fleece was in the temple of Zeus at Orchomenus, where Jason hung it on his return to Greece.

The Greeks knew the constellation as Κριός (Krios), meaning ram; Aries is the Latin name. On old star maps the ram is shown in a crouching position, but without wings, its head turned back towards Taurus. In the sky it is not at all prominent. Its most noticeable feature is a crooked line of three stars, which marks its head. Of these three stars, Alpha Arietis is called Hamal, from the Arabic for lamb; Beta Arietis is Sheratan, from the Arabic meaning 'two' of something (possibly two signs or two horns, for it was originally applied to both this star and to its neighbour, Gamma Arietis); and Gamma Arietis is Mesarthim, a curiously corrupted form of *al-sharatan*, the title which it originally shared with Beta Arietis.

First point of Aries

In astronomy, Aries assumes a far greater importance than its brightness would suggest, for in Greek times it contained the cardinal point known as the vernal equinox. This is the point at which the Sun crosses the celestial equator from north to south. But the vernal equinox is not stationary, because of the slow wobble of the Earth's axis known as precession.

When the Greek astronomer Hipparchus defined the position of the vernal equinox around 130 BC this point lay south of the star Mesarthim (Gamma Arietis). The zodiac was then taken to start from here, and so the vernal equinox was commonly known as the first point of Aries. Because of precession, the vernal equinox has moved some 30° since the time of Hipparchus and currently lies in the neighbouring constellation Pisces. By the year 2600 it will have entered Aquarius. Despite this, the vernal equinox is still sometimes called the first point of Aries.

Auriga
The charioteer

Genitive: Aurigae
Abbreviation: Aur
Size ranking: 21st
Origin: One of the 48 Greek constellations listed by Ptolemy in the *Almagest*
Greek name: Ἡνίοχος (Heniochos)

High in the northern sky stands a forlorn-looking charioteer. With his right hand he grasps the reins of a chariot, while on his left arm he carries a goat and its two kids. Of his chariot itself there is no sign. What's the story here? Mythology offers several identifications for this prominent constellation, although the presence of the goat is not accounted for by any of them.

The most popular interpretation is that he is Erichthonius, a legendary king of Athens. Erichthonius was the son of Hephaestus the god of fire, better known by his Roman name of Vulcan. Hephaestus was too busy smithying to be bothered with his son, who was instead raised by the goddess Athene, after whom the city of Athens is named. When he grew up, Erichthonius instituted a festival called the Panathenaea in her honour.

Athene taught Erichthonius many skills, including how to tame horses. He became the first person to harness four horses to a chariot, in imitation of the four-horse chariot of the Sun (the quadriga), a bold move which earned him the admiration of Zeus and assured him a place among the stars. There, according to this story, Erichthonius is depicted at the reins, perhaps participating in the Panathenaic games in which he frequently drove his chariot to victory.

Another identification is that Auriga is really Myrtilus, the charioteer of King Oenomaus of Pisa and son of Hermes. The king had a beautiful daughter, Hippodamia, whom he was determined not to let go. He challenged each of her suitors to a death-or-glory chariot race. They were to speed away with Hippodamia on their chariots, but if Oenomaus caught up with them before they reached Corinth he would kill them. Since he had the swiftest chariot in Greece, skilfully driven by Myrtilus, no man had yet survived the test.

A dozen suitors had been beheaded by the time that Pelops, the handsome son of Tantalus, came to claim Hippodamia's hand. Hippodamia, falling in love with him on sight, begged Myrtilus to betray the king so that Pelops might win the race. Myrtilus, who was himself secretly in love with Hippodamia, tampered with the pins holding the wheels on Oenomaus's chariot. During the pursuit of Pelops, the wheels of the king's chariot fell off and Oenomaus was thrown to his death.

Hippodamia was now left in the company of both Pelops and Myrtilus. Pelops solved the awkward situation by unceremoniously casting Myrtilus into the sea, from where he cursed the house of Pelops as he drowned. Hermes put the image of his son Myrtilus into the sky as the constellation Auriga. Germanicus Caesar supports this identification because, he says, 'you will

Auriga carrying the goat and kids, seen on Chart V of the *Uranographia* of Johann Bode (1801).The bright star Capella lies in the body of the goat. Its two kids are cradled on the charioteer's forearm. (Deutsches Museum)

observe that he has no chariot, and, his reins broken, is sorrowful, grieving that Hippodamia has been taken away by the treachery of Pelops'.

A third identification of Auriga is Hippolytus, son of Theseus, whose stepmother Phaedra fell in love with him. When Hippolytus rejected her, she hanged herself in despair. Theseus banished Hippolytus from Athens. As he drove away his chariot was wrecked, killing him. Asclepius the healer brought the blameless Hippolytus back to life again, a deed for which Zeus struck Asclepius down with a thunderbolt at the demand of Hades, who was annoyed at losing a valuable soul.

Aratus did not identify the constellation with any character. He simply called it Ἡνίοχος (Heniochos), the charioteer, as did Ptolemy in the *Almagest.* From this

Greek name comes the Latin transliteration Heniochus, used for the constellation by some Roman writers such as Manilius.

The she-goat and kids

Auriga contains the sixth-brightest star in the sky, known as Capella, a Roman name meaning 'she-goat'; the Greeks called it Αἴξ (Aix), meaning the same. Ptolemy in the *Almagest* described this star as being on the charioteer's left shoulder, but all major star atlases, including those of Bayer, Flamsteed, and Bode (previous page), have shown it in the body of the goat. According to Aratus it represented the goat Amaltheia, who suckled the infant Zeus on the island of Crete and was placed in the sky as a mark of gratitude, along with the two kids she bore at the same time.

The kids, frequently known by their Latin name of Haedi (Greek: Ἔριφοι, i.e. Eriphoi), are represented by the neighbouring stars Eta and Zeta Aurigae; Ptolemy described them as lying on the charioteer's left wrist. Hyginus credits the Greek astronomer Cleostratus with having first called these two stars the kids in the 5th century BC. It is sometimes said that the variable star Epsilon Aurigae to the north of them is a third member of the kids, but this is incorrect; Ptolemy and the mythologists were clear that there were only two kids. According to Ptolemy, Epsilon Aurigae marks the charioteer's left elbow.

An alternative story is that Amaltheia was not the goat itself but the nymph who owned the goat. Eratosthenes says that the goat was so ugly that it terrified the Titans who ruled the Earth at that time. When Zeus grew up he challenged the Titans for supremacy. Following the advice of an oracle he skinned the goat and made a cloak from its impenetrable hide, the back of which resembled the head of a Gorgon. This horrible-looking goatskin formed the so-called aegis of Zeus (the word aegis means 'goatskin'). The aegis protected Zeus and scared his enemies, a particular advantage in his fight against the Titans. Afterwards Zeus covered the bones of the goat in a normal-looking skin and transformed it into the star Capella.

Some early writers spoke of the Goat and Kids as a separate constellation, but since the time of Ptolemy they have been awkwardly combined with the charioteer, the goat resting on the charioteer's shoulder, with the kids supported on his forearm. There is no legend to explain why the charioteer is so encumbered with livestock.

A 'shared' star

Greek astronomers regarded one star as being shared by Auriga and Taurus. Old star maps show this star as representing both the right foot of the charioteer and also the tip of the bull's left horn. When the German astronomer Johann Bayer came to allocate Greek letters to the stars in the early 17th century he designated this star as both Gamma Aurigae and Beta Tauri. However, since the introduction of precise constellation boundaries in 1930, astronomers have assigned this star exclusively to Taurus as Beta Tauri and there is no longer a Gamma Aurigae. Hence, under the modern scheme, the bull has kept the tip of his horn but the luckless charioteer has lost his right foot.

Boötes

The herdsman

Genitive: Boötis
Abbreviation: Boo
Size ranking: 13th
Origin: One of the 48 Greek constellations listed by Ptolemy in the *Almagest*
Greek name: Βοώτης (Boötes)

This constellation (pronounced Boh-oh-tease) is closely linked in legend with the Great Bear, Ursa Major, because of its position behind the bear's tail. The origin of the name Boötes (Greek: Βοώτης) is not certain, but it probably comes from a Greek word meaning 'noisy' or 'clamorous', referring to the herdsman's shouts to his animals. An alternative explanation is that the name comes from the ancient Greek meaning 'ox-driver', from the fact that Ursa Major was sometimes visualized as a cart pulled by oxen.

The Greeks also knew this constellation as Ἀρκτοφύλαξ (Arctophylax), variously translated as Bear Watcher, Bear Keeper, or Bear Guard. Aratus wrote of 'Arctophylax, whom men also know as Boötes', and likened him to a man driving the bear around the pole. Homer in the *Odyssey* called him only Boötes, which suggests it is the older of the two alternative names. Later astronomers have given Boötes two dogs, in the form of the neighbouring constellation Canes Venatici, but they were not part of the original Greek visualization or legend.

According to a story that goes back to Eratosthenes, the constellation represents Arcas, son of the god Zeus and his paramour Callisto, daughter of King Lycaon of Arcadia. One day Zeus came to dine with the king, an unusual thing for a god to do. To test whether his guest really was the great Zeus, Lycaon cut up Arcas and served him as part of a mixed grill (some say that this deed was done not by Lycaon but by his sons). Zeus easily recognized the flesh of his own son. In a burning rage, he tipped over the table, scattering the feast, killed the sons of Lycaon with a thunderbolt, and turned Lycaon into a wolf. Then Zeus collected the parts of Arcas, made them whole again and gave his reconstituted son to Maia the Pleiad to raise.

Meanwhile, Callisto had been turned into a bear, some say by Zeus's wife Hera out of jealousy, or by Zeus himself to disguise his paramour from Hera's revenge, or even by Artemis to punish Callisto for losing her virginity. Whatever the case, when Arcas had grown into a strapping teenager he came across this bear while hunting in the woods. Callisto recognized her son, but though she tried to greet him warmly she could only growl. Not surprisingly, Arcas failed to interpret this expression of motherly love and began to chase the bear. With Arcas in hot pursuit, Callisto fled into the temple of Zeus, a forbidden place where trespassers were punished by death. Zeus snatched up Arcas and his mother and placed them in the sky as the constellations of the bear-keeper and the bear.

Boötes on Chart VII of the *Uranographia* of Johann Bode (1801). He carries a club or staff in his right hand and a sickle in his left, with which he also grasps the leash of his hunting dogs, represented by neighbouring Canes Venatici. Here Boötes is standing on Mons Maenalus, an obsolete sub-constellation (page 198). Above his head is another obsolete constellation, Quadrans Muralis (page 200). (Deutsches Museum)

A second legend identifies Boötes with Icarius (not to be confused with Icarus, son of Daedalus). According to this gloomy tale, recounted at length by Hyginus in *Poetic Astronomy*, the god Dionysus taught Icarius how to cultivate vines and make wine. When he offered some of his new vintage to shepherds they became so intoxicated that their friends thought Icarius had poisoned them, and in revenge they killed him.

His dog Maera fled home howling and led Icarius's daughter Erigone to where his body lay beneath a tree. In despair, Erigone hanged herself from the tree; even the dog died, either of grief or by drowning itself. Zeus put Icarius into the sky as Boötes, his daughter Erigone became the constellation Virgo and the dog became Canis Minor or, in some versions, Canis Major.

Boötes contains the fourth-brightest star in the entire sky, Arcturus (Ἀρκτοῦρος in Greek), mentioned by Homer, Hesiod, Aratus, and Ptolemy. The name means 'bear guard'. Aratus described it as lying beneath his belt, while Germanicus Caesar said it 'lies where his garment is fastened by a knot'. Ptolemy placed it between the thighs, which is where mapmakers have traditionally depicted it. To the eye, Arcturus has a noticeably orange colour. Astronomers have found that it is a red giant star about 25 times larger than the Sun, lying 37 light years away.

Caelum

The chisel

Genitive: Caeli
Abbreviation: Cae
Size ranking: 81st
Origin: The 14 southern constellations of Nicolas Louis de Lacaille

This small and insignificant constellation in the southern hemisphere, representing an engraver's chisel, is one of the inventions of the 18th-century French astronomer Nicolas Louis de Lacaille. He introduced it on his map of the southern stars published in 1756 under the French name les Burins, which he Latinized to Caelum Scalptorium on the second edition of the map in 1763. In

Caelum shown under the name Caela Scalptoris on Chart XVIII of the *Uranographia* of Johann Bode (1801), who added two scribing tools to the burin and échoppe described by its inventor, Lacaille. (Deutsches Museum)

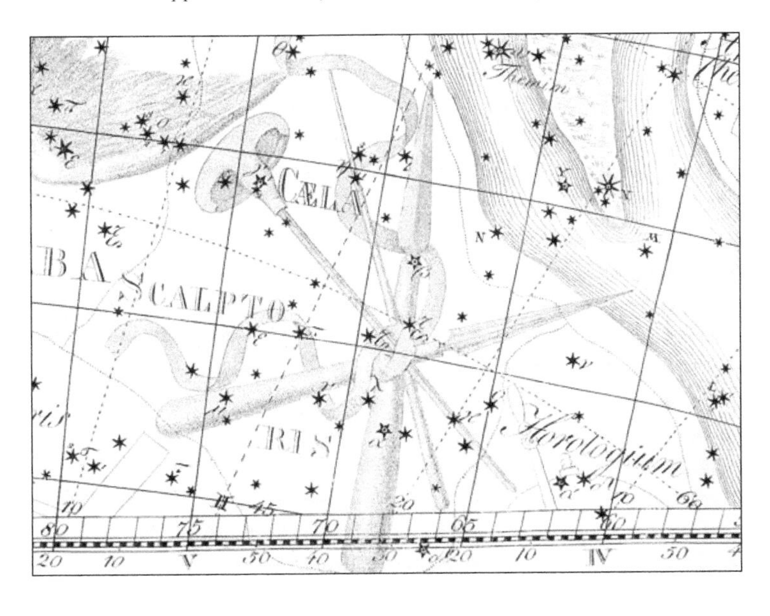

1844 the English astronomer John Herschel proposed shortening it to Caelum. Francis Baily adopted this suggestion in his *British Association Catalogue* of 1845, and it has been known as Caelum ever since.

Lacaille said that the constellation represented two engraving tools, crossed and connected by a ribbon. One tool was a burin, a sharp-tipped cold chisel also known as a graver, while the other was an échoppe, a type of etching needle invented by the 17th-century French printmaker Jacques Callot. It is now just referred to as the chisel. There are no legends associated with the constellation and its stars are faint, of 4th magnitude and below.

Camelopardalis
The giraffe

Genitive: Camelopardalis
Abbreviation: Cam
Size ranking: 18th
Origin: Petrus Plancius

One of the most unlikely animals to be found in the sky is a giraffe. The Greeks called giraffes 'camel leopards' because of their long necks and spots, which is where the name Camelopardalis comes from. However, the constellation Camelopardalis was not invented by the Greeks but by the Dutch theologian

The top half of Camelopardalis, shown on Chart III of the *Uranographia* atlas of Johann Bode (1801). Also included are the now-obsolete constellations of Rangifer, the reindeer, and Custos Messium, the harvest keeper – see Chapter Four. (Deutsches Museum)

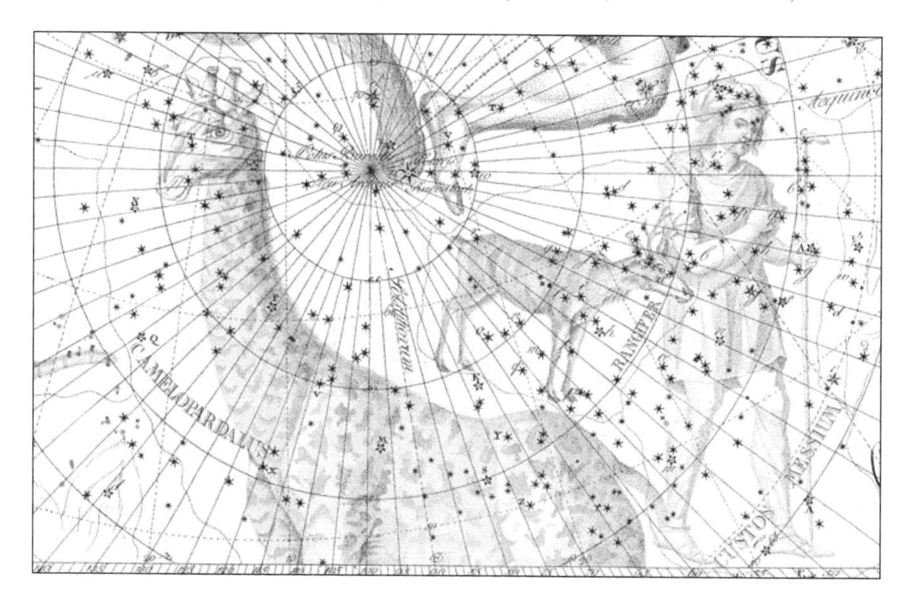

and astronomer Petrus Plancius in 1612. Plancius first showed it on a celestial globe in that year along with another odd constellation that is still recognized, Monoceros. Plancius used the same name for the constellation as we do, but on some old maps the name is also written as Camelopardalus or Camelopardus. Camelopardalis lies in the far northern sky between the head of the Great Bear and Cassiopeia, an area that was left blank by the Greeks because it contains no stars brighter than fourth magnitude.

The exact significance of the constellation is unclear. The German astronomer Jacob Bartsch (*c.*1600–33) included Camelopardalis on his map of 1624, which was its first appearance in print as distinct from its depiction on a globe. Bartsch interpreted it as the camel on which Rebecca rode into Canaan for her marriage to Isaac, as told in Chapter 24 of the book of Genesis. But Camelopardalis is a giraffe not a camel, so Bartsch's explanation is unsatisfactory. Bartsch seems not to have known much about this constellation, for he wrongly attributed its invention to Isaac Habrecht of Strasbourg, who had shown it on his star globe of 1621.

Its three brightest stars were labelled Alpha, Beta, and Gamma by the English astronomer Francis Baily in his *British Association Catalogue* of 1845. By modern measurements Beta is magnitude 4.0 and Alpha is 4.3, so this is one of the constellations in which Alpha is not the brightest star.

Cancer
The crab

Genitive: Cancri
Abbreviation: Cnc
Size ranking: 31st
Origin: One of the 48 Greek constellations listed by Ptolemy in the *Almagest*
Greek name: Καρκίνος (Karkinos)

The crab is a minor character in one of the greatest epics of Greek myth: the labours of Heracles (the Greek name for Hercules). While Heracles was fighting the multi-headed monster called the Hydra in the swamp near Lerna, the crab emerged from the swamp and added its own attack by biting Heracles on the foot. Heracles angrily stamped on the crab, crushing it. For this modest contribution to history, we are told that the goddess Hera, the enemy of Heracles, put the crab among the stars of the zodiac. Its name in Greek was Καρκίνος (Karkinos), or Carcinus in Latin transliteration.

Fittingly enough for such a minor character it is the faintest of the zodiacal constellations, with no star brighter than fourth magnitude. The star Alpha Cancri is named Acubens, from the Arabic meaning 'claw'. As described by Ptolemy in the *Almagest*, this star lies on the southern claw of the crab; the northern claw is marked by Iota Cancri. Beta and Mu Cancri lie on the southern and northern rear legs, respectively.

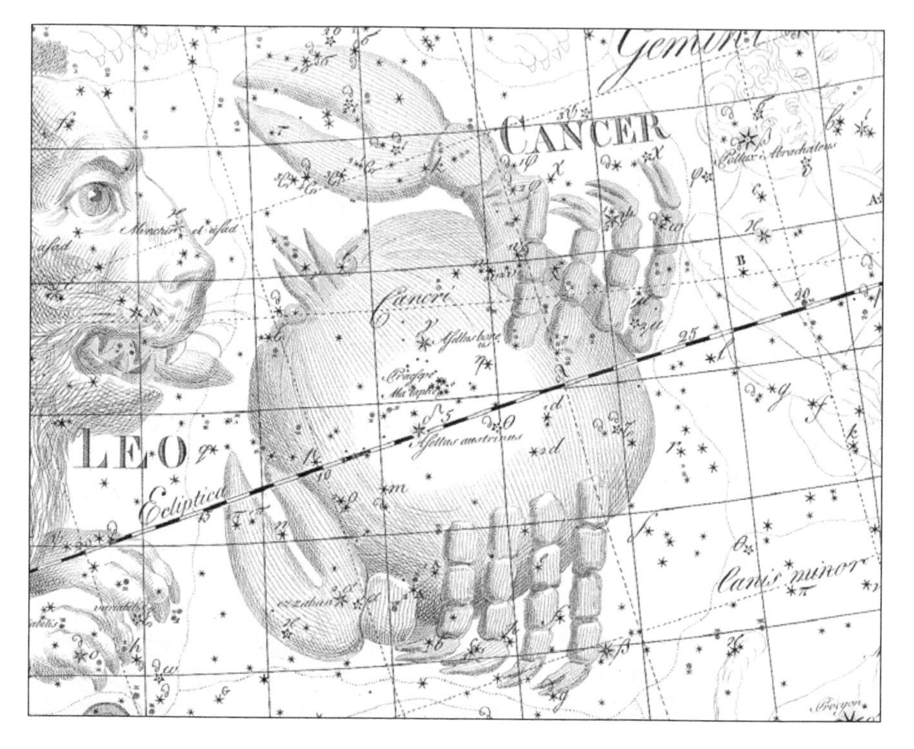

Cancer, from Chart XIII of Johann Bode's *Uranographia* (1801). At its centre lies the star cluster Praesepe, flanked to the north and south by the stars Asellus Borealis and Asellus Australis (Gamma and Delta Cancri). (Deutsches Museum)

The asses and the manger

Gamma and Delta Cancri were known to the Greeks as Ὄνοι (Onoi), the asses; we know them by their Latin names, Asellus Borealis and Asellus Australis, the northern ass and southern ass, and they have their own legend. According to Eratosthenes, during the battle between the gods and the Giants which followed the overthrow of the Titans, the gods Dionysus, Hephaestus, and some companions came riding on donkeys to join the fray. The Giants had never heard the braying of donkeys before and took flight at the noise, thinking that some dreadful monster was about to be unleashed upon them. Dionysus put the asses in the sky, either side of the cluster of stars which the Greeks called Φάτνη (Phatne), the manger, from which the asses seem to be feeding. Ptolemy described Phatne as 'the nebulous mass in the chest'. Astronomers now know this star cluster by its Latin name Praesepe, but it is popularly termed the Beehive – praesepe can mean both 'manger' and 'hive'.

The constellation gives its name to the tropic of Cancer, the latitude on Earth at which the Sun appears overhead at noon on the summer solstice, June 21. In ancient Greek times the Sun lay among the stars of Cancer on this date, but the wobble of the Earth on its axis called precession has since moved the summer solstice from Cancer through neighbouring Gemini and into Taurus.

Canes Venatici
The hunting dogs

Genitive: Canum Venaticorum
Abbreviation: CVn
Size ranking: 38th
Origin: The seven constellations of Johannes Hevelius

The Polish astronomer Johannes Hevelius formed this constellation in 1687 from a scattering of faint stars beneath the tail of Ursa Major. Canes Venatici represents a pair of hounds held on a lead by Boötes, snapping at the heels of the great bear. The constellation's two brightest stars, Alpha and Beta Canum Venaticorum, both lie in the southern dog.

Ptolemy had listed both these stars in the *Almagest* as among the 'unformed' stars outside the figure of the great bear, not belonging to any particular

Canes Venatici, a pair of hunting dogs held on a leash by Boötes, seen in the *Atlas Coelestis* of John Flamsteed (1729). (University of Michigan Library)

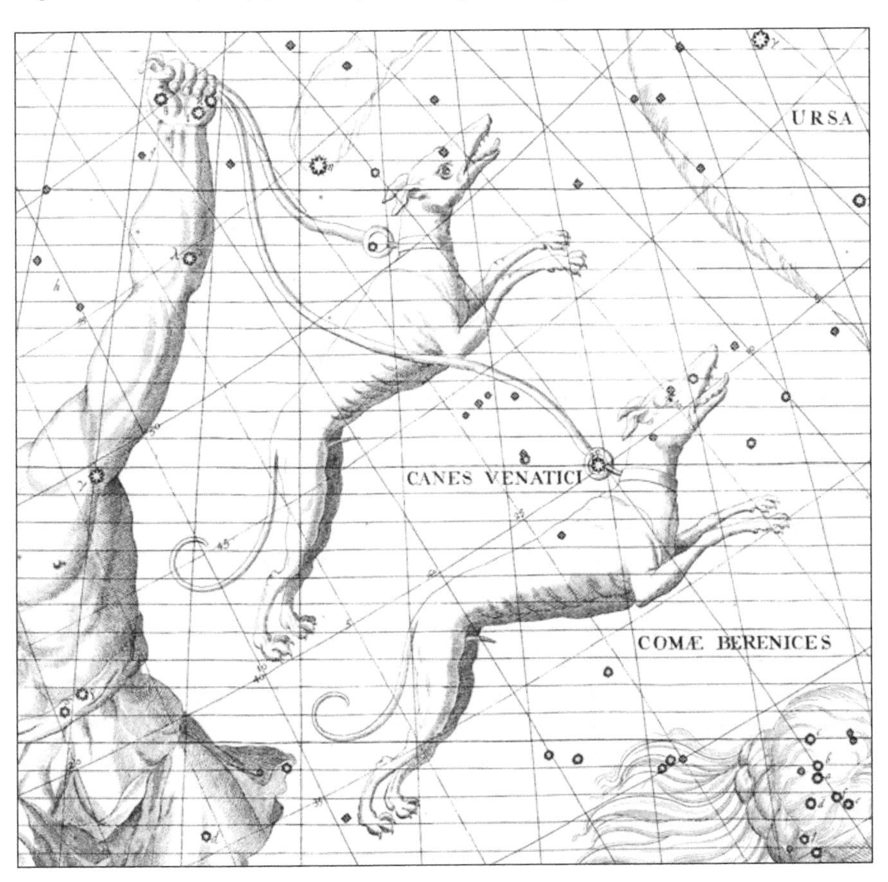

constellation, so they were free to be used in a new figure. Earlier in the 17th century the Dutch astronomer Petrus Plancius had introduced a new constellation called Jordanus, the river Jordan, that started in this area (see page 195), but Hevelius replaced it with three inventions of his own: Canes Venatici, Leo Minor, and Lynx.

The idea of dogs being held by Boötes was not original to Hevelius. On a star chart published in 1533 the German astronomer Peter Apian (1495–1552) showed Boötes with two dogs at his heels and holding their leash in his right hand. On another chart published by Apian three years later the number of dogs had grown to three and the leash had moved to the left hand, but they were still following Boötes and not the bear. In neither case was any attempt made to connect the dogs with charted stars, nor were they named, so the credit for showing the dogs in their current position and for making them a separate constellation remains with Hevelius.

Charles's Heart, and a Whirlpool

Where the southern dog's lead is attached to its collar lies the star Alpha Canum Venaticorum, known as Cor Caroli, meaning Charles's Heart, in honour of King Charles I of England who was executed by parliamentarians in 1649. The star was given this title by Charles Scarborough, physician to Charles's son, King Charles II, before Canes Venatici was formed. Scarborough reputedly said that this star shone particularly brightly on the night of 1660 May 29, when Charles II returned to London at the Restoration of the Monarchy. Because of this there has been much confusion over which King Charles the star is intended to commemorate, but it definitely refers to the first King Charles.

The name first appeared on a chart of 1673 by the English cartographer Francis Lamb (*fl.*1670–1700), who labelled it Cor Caroli Regis Martyris, a reference to the fact that King Charles I was beheaded (or 'martyred', as Lamb loyally put it). Lamb and others, such as the ardent royalist Edward Sherburne in 1675, drew a heart around the star surmounted by a crown, turning it into a mini-constellation. Johann Bode, in his *Uranographia* atlas of 1801, retained the heart and crown on the neck of the southern dog, but erred by calling the star Cor Caroli II, thinking that it referred to the second King Charles. Despite its overtly nationalistic nature, the name has stuck.

Beta Canum Venaticorum, on the dog's snout, is called Chara, from the Greek for 'joy', also the name that Hevelius gave the southern dog. The northern dog, which Hevelius called Asterion ('little star'), is marked only by a scattering of faint stars. Bode drew the dogs with their names engraved on their collars. Alpha and Beta Canum Venaticorum are the only two stars in the constellation with Greek letters, which they were given by the English astronomer Francis Baily in his *British Association Catalogue* of 1845.

Canes Venatici contains a beautiful spiral galaxy, M51, called the Whirlpool. M51 was the first galaxy in which spiral form was noticed, by the Irish astronomer Lord Rosse (1800–67) in 1845. It consists of a large galaxy about 25 million light years away in near-collision with a smaller one; it is thought that the two galaxies may eventually merge.

Canis Major

The greater dog

Genitive: Canis Majoris
Abbreviation: CMa
Size ranking: 43rd
Origin: One of the 48 Greek constellations listed by Ptolemy in the *Almagest*
Greek name: Κύων (Kyon)

Four dogs are to be found among the constellations: Canis Major, Canis Minor, and the two hunting dogs, Canes Venatici, but Canis Major is undoubtedly the top dog. Indeed, Ptolemy in the *Almagest* called it simply Κύων (Kyon), the dog. Canis Major is dominated by Sirius, popularly termed the Dog Star, the most brilliant star in the entire night sky; almost certainly the constellation originated with this star alone.

Aratus referred to Canis Major as the guard-dog of Orion, following on the heels of its master and standing on its hind legs with Sirius carried in its jaws. Manilius called it 'the dog with the blazing face' while Germanicus Caesar said that 'it belches forth fire from its mouth'. Canis Major seems to cross the sky in

Canis Major on Chart XVIII of Johann Bode's *Uranographia* (1801). Bode depicted the dog with Sirius marking its snout, whereas classical Greek descriptions placed Sirius in the dog's jaws. (Deutsches Museum)

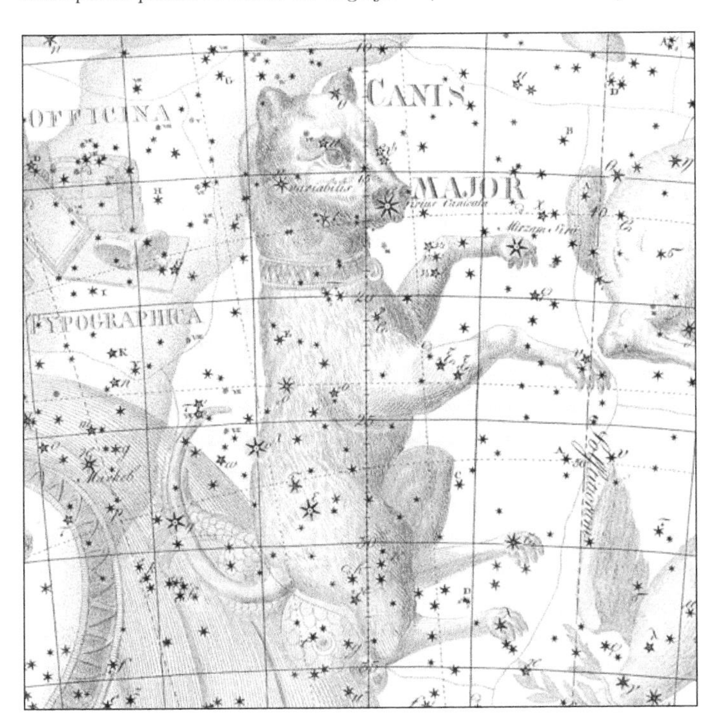

pursuit of the hare, represented by the constellation Lepus under Orion's feet. To the north of it scampers Canis Minor, the smaller dog, either having lost the scent or sniffing out different prey.

Mythologists such as Eratosthenes and Hyginus said that the constellation represented Laelaps, a dog so swift that no prey could outrun it. This dog had a long list of owners, one of them being Procris, daughter of King Erechtheus of Athens and wife of Cephalus, but accounts differ about how she came by it. In one version the dog was given to her by Artemis, goddess of hunting; but a more likely account says that it is the dog given by Zeus to Europa, whose son Minos, King of Crete, passed it on to Procris. The dog was presented to her along with a javelin that could never miss; this turned out to be an unlucky gift, for her husband Cephalus accidentally killed her with it while out hunting.

Cephalus inherited the dog, and took it with him to Thebes (not Thebes in Egypt but a town in Boeotia, north of Athens) where a vicious fox was ravaging the countryside. The fox was so swift of foot that it was destined never to be caught – yet Laelaps the hound was destined to catch whatever it pursued. Off they went, almost faster than the eye could follow, the inescapable dog in pursuit of the uncatchable fox. At one moment the dog would seem to have its prey within grasp, but could only close its jaws on thin air as the fox raced ahead of it again. There could be no resolution of such a paradox, so Zeus turned them both to stone, and the dog he placed in the sky as Canis Major, without the fox.

Sirius, the dazzling dog star

The name of the star Sirius comes from the Greek word Σείριος (seirios) meaning 'searing' or 'scorching', highly appropriate for something so brilliant. Even though the name Sirius was known as far back as the time of Hesiod (*c.*700 BC), Ptolemy in the *Almagest* called it Κύων (Kyon), 'the Dog', the same name as for the whole constellation. He described it as 'the star in the mouth'.

In Greek times the rising of Sirius at dawn just before the Sun marked the start of the hottest part of the summer, a time that hence became known as the Dog Days. 'It barks forth flame and doubles the burning heat of the Sun', said Manilius, expressing a belief held by the Greeks and Romans that the star had a heating effect. The ancient Greek writer Hesiod wrote of 'heads and limbs drained dry by Sirius', and Virgil in the *Georgics* said that 'the torrid Dog Star cracks the fields'. Germanicus Caesar claimed that when it rose with the Sun it strengthened healthy crops but killed those with shrivelled foliage or feeble roots. 'There is no star the farmer likes more or hates more', according to Germanicus.

'Hardly is it inferior to the Sun, save that its abode is far away', wrote Manilius, anticipating the modern view that stars are bodies like the Sun only vastly more distant. Yet, in contradiction of the supposed heating effects of Sirius, Manilius continued: 'The beams it launches from its sky-blue face are cold'. That description of the colour of Sirius is in contrast to Ptolemy's surprising reference to it as reddish, which has caused all manner of arguments. In fact, Manilius was nearly correct, for Sirius is a blue-white star, even larger and brighter than the Sun. It lies 8.6 light years away, making it one of the Sun's closest neighbours. It has a white dwarf companion star, visible only through telescopes, that orbits it every 50 years.

In 14th-century Europe, Sirius was also known as Alhabor or Alabor, an Arabic name that can be found on astrolabes of the time. Geoffrey Chaucer used the name Alhabor for Sirius in his celebrated *Treatise on the Astrolabe* written in or around 1391. However, astronomers eventually settled on the original Greek name in preference to the Arabic alternative.

Beta Canis Majoris, which precedes Sirius across the sky, is known as Mirzam, from the Arabic *al-mirzam*. According to the 10th-century Arabic astronomer al-Ṣūfī the Arabs gave this name to any star that preceded a bright star. Hence it was also applied to Beta Canis Minoris, which precedes Procyon, and Gamma Orionis, which precedes Betelgeuse, but it is the attribution to Beta Canis Majoris that has stuck.

Ptolemy listed 11 stars as lying around the constellation but not forming part of it. Of these, nine were later used by Petrus Plancius to create a new constellation, Columba, the dove; one star was transferred to Monoceros; and one was eventually incorporated in Canis Major.

Canis Minor

The lesser dog

Genitive: Canis Minoris
Abbreviation: CMi
Size ranking: 71st
Origin: One of the 48 Greek constellations listed by Ptolemy in the *Almagest*
Greek name: Προκύων (Prokyon)

Representing the smaller of the two dogs of Orion, Canis Minor originally consisted of just the bright star Procyon, known in Greek as Προκύων (Prokyon), meaning 'before the dog' or 'foredog'. This name, used by the Greeks for both the star and the constellation, comes from the fact that it rises earlier than its more prominent kennel-mate Canis Major which Ptolemy called simply Κύων (Kyon), the Dog. An alternative Latin name for Canis Minor was Antecanis or Anticanis, also meaning 'before the dog'.

Canis Minor is small and contains little of interest other than Procyon itself, the eighth-brightest star in the heavens, 11.5 light years away. Ptolemy in the *Almagest* catalogued just two stars in Canis Minor: Procyon in the dog's body, and the star in the dog's neck now known as Beta Canis Minoris, also called Gomeisa. Procyon is of particular interest to astronomers because it has a small, hot companion star called a white dwarf that orbits it every 41 years. Coincidentally the other dog star, Sirius, also has one of these small, highly dense white dwarfs as a companion.

Canis Minor is usually identified as one of the dogs of Orion, following him across the sky as the Earth turns. But in a famous legend from Attica (the area

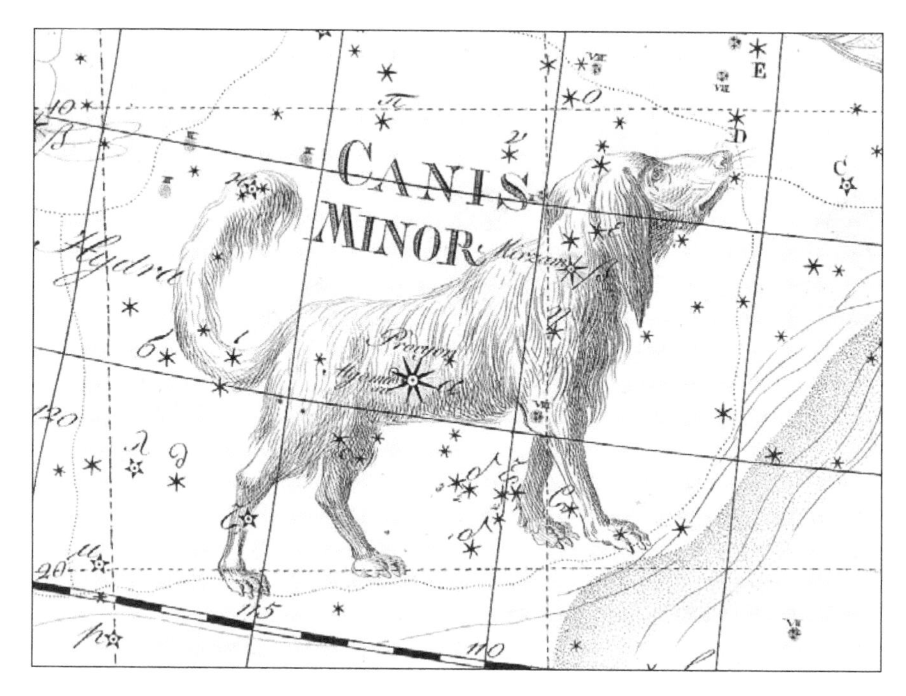

Canis Minor seen on Chart XII of the *Uranographia* of Johann Bode (1801). In its body lies the bright star Procyon. On this chart Bode labels Beta Canis Minoris in the dog's neck as Mirzam, an Arabic title now applied to Beta Canis Majoris. (Deutsches Museum)

around Athens), recounted by the mythographer Hyginus, the constellation represents Maera, dog of Icarius, the man whom the god Dionysus first taught to make wine. When Icarius gave his wine to some shepherds for tasting, they rapidly became drunk. Suspecting that Icarius had poisoned them, they killed him. Maera the dog ran howling to Icarius's daughter Erigone, caught hold of her dress with his teeth and led her to her father's body. Both Erigone and the dog took their own lives where Icarius lay.

Zeus placed their images among the stars as a reminder of the unfortunate affair. To atone for their tragic mistake, the people of Athens instituted a yearly celebration in honour of Icarius and Erigone. In this story, Icarius is identified with the constellation Boötes, Erigone is Virgo, and Maera is Canis Minor.

According to Hyginus, the murderers of Icarius fled to the island of Ceos off the coast of Attica, but their wrongdoing followed them. The island was plagued with famine and sickness, attributed in the legend to the scorching effect of the Dog Star (here, Procyon seems to have become confused with the greater dog star, Sirius in Canis Major). King Aristaeus of Ceos, son of the god Apollo, asked his father for advice and was told to pray to Zeus for relief. Zeus sent the Etesian winds, which every year blow for 40 days from the rising of the Dog Star to cool all of Greece and its islands in the summer heat. After this, the priests of Ceos instituted the practice of making yearly sacrifices before the rising of the Dog Star.

Capricornus

The sea goat

Genitive: Capricorni
Abbreviation: Cap
Size ranking: 40th
Origin: One of the 48 Greek constellations listed by Ptolemy in the *Almagest*
Greek name: Αἰγόκερως (Aigokeros)

Capricornus is an unlikely looking creature, with the head and forelegs of a goat and the tail of a fish. The constellation evidently originated with the Sumerians and Babylonians, who had a fondness for amphibious creatures; the ancient Sumerians called it SUHUR-MASH-HA, the goat-fish. But to the Greeks, who named it Αἰγόκερως (Aigokeros, or Aegoceros in Latin transliteration), meaning goat-horned, the constellation was identified with Pan, god of the countryside, who had the horns and legs of a goat.

Pan, a playful creature of uncertain parentage, spent much of his time chasing females or sleeping it off with a siesta. He could frighten people with his loud shout, which is the origin of the word 'panic'. One of his offspring was Crotus, identified with the constellation Sagittarius. Pan's attempted seduction

Capricornus as shown on Chart XVI of the *Uranographia* of Johann Bode (1801). Beneath it floats the now-abandoned constellation of Globus Aerostaticus, the hot-air balloon (see page 193). The dashed line is the Sun's annual path, the ecliptic. (Deutsches Museum)

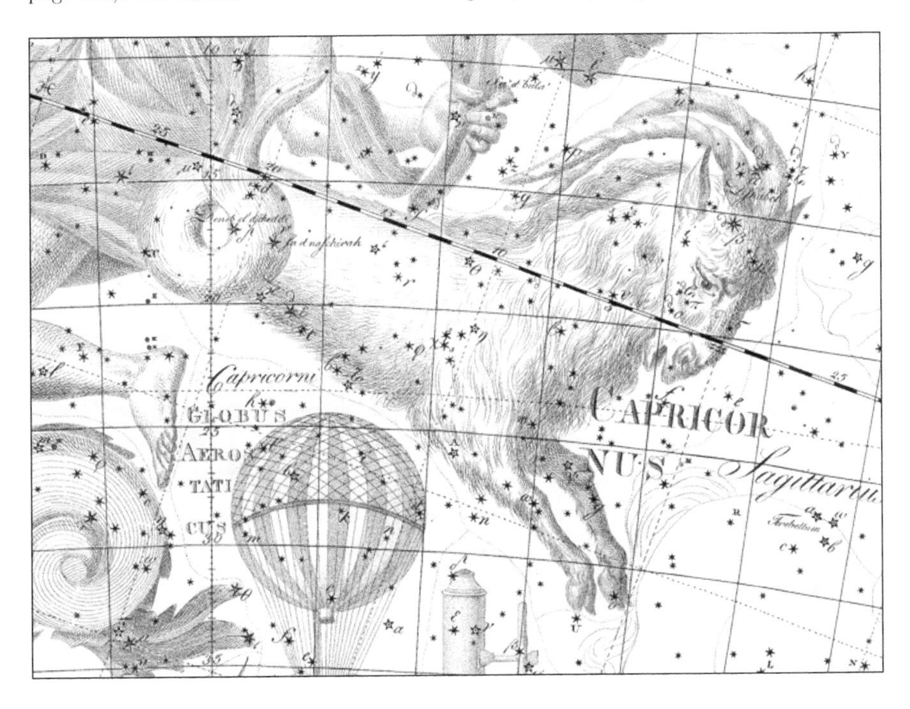

of the nymph Syrinx failed when she turned herself into a handful of reeds. As he clutched the reeds the wind blew through them, creating an enchanting sound. Pan selected reeds of different lengths and stuck them together with wax to form the famous pipes of Pan, also called the syrinx.

Pan came to the rescue of the gods on two separate occasions. During the battle of the gods and the Titans, Pan blew a conch shell to help put the enemy to flight. According to Eratosthenes his connection with the conch shell accounts for his fishy nature in the sky, although Hyginus says somewhat absurdly that it is because he hurled shellfish at the enemy. On a later occasion, Pan shouted a warning to the gods that the monster Typhon was approaching, sent by Mother Earth (Gaia) against the gods. At Pan's suggestion the gods disguised themselves as animals to elude the monster. Pan himself took refuge in a river, turning the lower part of his body into a fish.

Zeus grappled with Typhon, but the monster pulled out the sinews from Zeus's hands and feet, leaving the god crippled. Hermes and Pan replaced the sinews, allowing Zeus to resume his pursuit of Typhon. Zeus cut down the monster with thunderbolts and finally buried him under Mount Etna in Sicily, which still belches fire from the monster's breath. In gratitude for these services, Zeus placed the image of Pan in the sky as the constellation Capricornus.

The star Alpha Capricorni is called Algedi, from the Arabic *al-jady* meaning 'the kid', the Arabic name for the constellation. Delta Capricorni is called Deneb Algedi, from the Arabic for 'the kid's tail'. As defined by the modern IAU boundaries Capricornus is the smallest of the 12 constellations of the zodiac, occupying less than a third the area of the largest zodiacal constellation, Virgo.

The constellation gives its name to the tropic of Capricorn, the latitude on Earth at which the Sun appears overhead at noon on the winter solstice, around December 22. In Greek times the Sun was in Capricornus on this date, but as a result of precession the Sun is now in Sagittarius at the winter solstice.

Carina

The keel

Genitive: Carinae
Abbreviation: Car
Size ranking: 34th
Origin: Part of the original Greek constellation Argo Navis

The smallest but most prominent of the three parts into which the ancient Greek constellation of Argo Navis, the ship of the Argonauts, was divided by the French astronomer Nicolas Louis de Lacaille in his first catalogue of the southern stars, published in 1756. In that catalogue he gave it the French name Corps du Navire. His final catalogue of 1763 contained the same subdivisions but with Latin instead of French names. The other two parts are Puppis, the poop or stern, and Vela, the sails. For the full story of Argo, and an illustration, see page 187. Although usually described as the keel, Carina represents the main body or hull of the ship. It contains the second-brightest star in the entire sky, Canopus, a creamy white giant just over 300 light years away, which marks the blade of one of the ship's two steering oars. Eratosthenes and Ptolemy both spelled the star's name Κάνωβος (Kanobos); Canopus is the Latinized version.

Canopus was not mentioned by Aratus, because the star was below the horizon from Greece in his day. The name first appears with Eratosthenes who was based farther south, at Alexandria in northern Egypt. From there he could see it low in the south, as could Ptolemy, who worked at Alexandria four centuries later. It was the most southerly star that Ptolemy catalogued in his *Almagest*.

Greek writers such as Conon (*c.*280–*c.*220 BC) and Strabo (*c.*64 BC–*c.* AD 24) tell us that Canopus is named after the helmsman of the Greek King Menelaus. On Menelaus's return from Troy with Helen his fleet was driven off-course by a storm and landed in Egypt. There Canopus died of a snake bite; Helen killed the snake, and she and Menelaus buried Canopus with full honours. On that site grew the city of Canopus (the modern Abu Qir) at the mouth of the Nile. Fittingly, modern space probes now use Canopus as a navigation star. Eratosthenes also knew this star by the name Περίγειος (i.e. Perigeios, or Perigee), in reference to the fact that it remained close to the horizon; this name appeared in Eratosthenes's entry on Eridanus, not Argo.

The constellation contains a unique star, Eta Carinae, which flared up to become brighter than Canopus in 1843, but has since faded to the edge of naked-eye visibility. Astronomers now think that Eta Carinae consists of a close pair of hot, very massive stars, one or both of which will one day explode as a supernova. The second-magnitude stars Epsilon and Iota Carinae, along with Delta and Kappa Velorum to the north in Vela, form a cruciform shape known as the False Cross, sometimes mistaken for the true Southern Cross.

Cassiopeia

Genitive: Cassiopeiae
Abbreviation: Cas
Size ranking: 25th
Origin: One of the 48 Greek constellations listed by Ptolemy in the *Almagest*
Greek name: Κασσιέπεια (Kassiepeia)

Cassiopeia was the vain and boastful wife of King Cepheus of Ethiopia, who stands next to her in the sky. They are the only husband-and-wife couple among the constellations. Classical authors spelled her name Cassiepeia, from the original Greek Κασσιέπεια, but Cassiopeia is the form used by astronomers.

While combing her long locks one day, Cassiopeia dared to claim that she was more beautiful than the sea nymphs called the Nereids. Such hubris by a mortal could not go unpunished so the Nereids went in search of retribution. There were 50 Nereids, all daughters of Nereus, the so-called Old Man of the Sea, and one of them, Amphitrite, was married to Poseidon, the sea god.

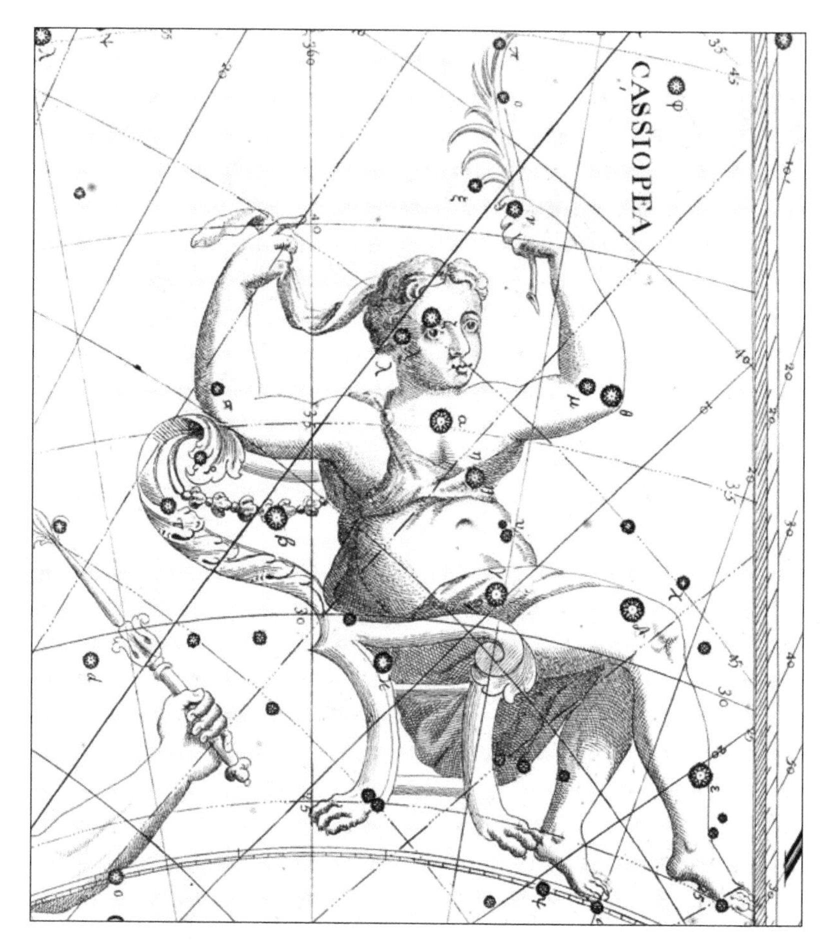

Cassiopeia the vainglorious queen seated on her throne, depicted in the *Atlas Coelestis* of John Flamsteed (1729). (University of Michigan Library)

Amphitrite and her sisters appealed to Poseidon to punish Cassiopeia for her vanity. Bowing to their request, the sea god sent a monster to ravage the coast of King Cepheus's country. This monster is commemorated in the constellation Cetus (page 73). To appease the monster, Cepheus and Cassiopeia chained their daughter Andromeda to a rock as a sacrifice, but Andromeda was saved from the monster's jaws by the hero Perseus in one of the most famous rescue stories in history (pages 37–38).

In the sky, Cassiopeia is depicted sitting on her throne. Each night she circles the celestial pole, sometimes upright, sometimes hanging upside down in apparent danger of falling out. The mythologists interpreted the indignity of this celestial fairground ride as part of her punishment from the gods, who made her a figure of fun. Aratus wrote that she plunged headlong into the sea like a diver (some translate it as 'tumbler'), her feet waving in the air, because as seen from Greek latitudes she would have received a ducking at the lowest point on each circuit. Her long-suffering husband Cepheus alongside her endured the same fate.

Germanicus Caesar described Cassiopeia thus: 'Her face contorted in agony, she stretches out her hands as if bewailing abandoned Andromeda, unjustly atoning for the sin of her mother', and this is how she is drawn in early manuscripts illustrating the works of Aratus and Hyginus. However, from the time of Dürer onwards she was portrayed not with her arms outstretched but holding aloft a palm frond in one hand. With her other hand she is either holding a robe or fussing with her hair, as in the illustration opposite.

The five brightest stars of Cassiopeia are arranged in a distinctive W-shape which Aratus and other writers likened to a key or a folding door. Alpha Cassiopeiae is called Schedar, from the Arabic *al-sadr* meaning 'the breast', where Ptolemy said it lay. Beta Cassiopeiae is known as Caph from the Arabic meaning 'stained hand', because the stars of Cassiopeia were thought by the Arabs to represent a hand tattooed with henna. Delta Cassiopeiae is named Ruchbah, from the Arabic for 'knee', *rukbat.* The central star of the W shape, Gamma Cassiopeiae, is an erratic variable star, given to occasional outbursts in brightness; it has no official name.

Tycho's Star

In November 1572 the familiar W-shape of Cassiopeia was disturbed by a bright interloper, now called Tycho's Star after the Danish astronomer Tycho Brahe who first spotted it. Tycho published a treatise about the 'new' star the following year, *De nova et nullius ævi memoria prius visa stella.* We now know that the interloper was actually a supernova, the explosion of a massive star at the end of its life. At its brightest, Supernova 1572 reached magnitude −4, comparable to Venus. It remained visible to the naked eye for over a year.

Centaurus

The centaur

Genitive: Centauri
Abbreviation: Cen
Size ranking: 9th
Origin: One of the 48 Greek constellations listed by Ptolemy in the *Almagest*
Greek name: Κένταυσος (Kentauros)

Centaurs were mythical beasts, half-man, half-horse. They were a wild and ill-behaved race, particularly when the wine bottle was opened. But one centaur, Chiron, stood out from the rest as being wise and scholarly, and he is the one who is represented by the constellation Centaurus (Κένταυσος in Greek).

Chiron was born of different parents from the other centaurs, which accounts for his difference in character. His father was Cronus, king of the Titans, who one day caught and seduced the sea nymph Philyra. Surprised in

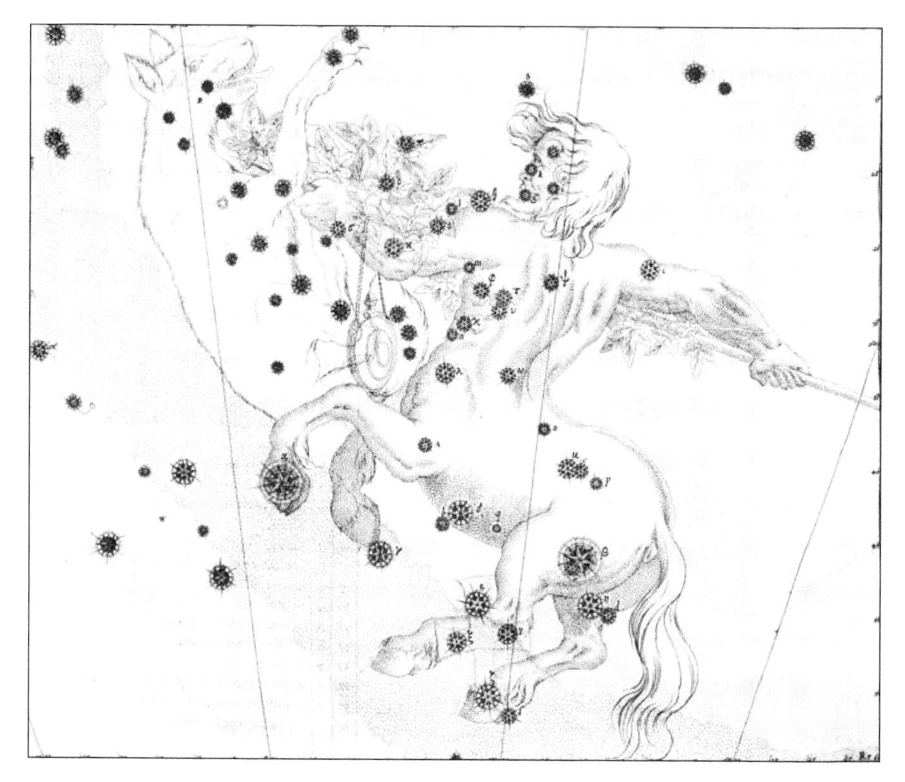

Centaurus from the *Uranometria* of Johann Bayer (1603). The centaur holds a long pole called a thyrsus on which is impaled Lupus, the wolf. Alpha Centauri, the closest star to the Sun, marks the centaur's forefoot which splashes in the Milky Way. Here Bayer has overdrawn the Southern Cross on the centaur's hind legs. (Linda Hall Library)

the act by his wife Rhea, Cronus turned himself into a horse and galloped away, leaving Philyra to bear a hybrid son.

Chiron grew up to be a skilled teacher of hunting, medicine, and music; his cave on Mount Pelion in eastern Greece became a veritable academy for young princes in search of a good education. Chiron was so trusted by the gods and heroes of ancient Greece that he was made foster-father to Jason (of Argonauts fame) and Achilles; but perhaps his most successful pupil was Asclepius, son of Apollo, who became the greatest of all healers and is commemorated in the constellation Ophiuchus.

For a creature who did so much good during his lifetime, Chiron suffered a tragic death. It arose from a visit paid by Heracles to the centaur Pholus, who entertained him to dinner and offered him wine from the centaurs' communal jar. When the other centaurs realized their wine was being drunk they burst angrily into the cave, armed with rocks and trees. Heracles repulsed them with a volley of arrows. Some of the centaurs took refuge with Chiron, who had been innocent of the attack, and an arrow of Heracles accidentally struck Chiron in the knee. Heracles, concerned for the good centaur, pulled out the arrow, apologizing profusely, but he already knew that Chiron was doomed. Even Chiron's best medicine was no match for the poison of the Hydra's blood in which Heracles had dipped his arrows.

Aching with pain, but unable to die because he was the immortal son of Cronus, Chiron retreated to his cave. Rather than let him suffer endlessly, Zeus agreed that Chiron should transfer his immortality to Prometheus. Thus released, Chiron died and was placed among the stars. Another version of the story simply says that Heracles visited Chiron and that while the two were examining his arrows one accidentally dropped on the centaur's foot.

In the sky, the centaur is visualized as about to sacrifice a wild animal (the adjoining constellation Lupus) on the altar, Ara. Eratosthenes says that this is a sign of Chiron's virtue. Eratosthenes also offers an alternative scene: in this, the wild animal is instead a wine skin, from which the centaur is about to pour an offering onto the altar. The centaur holds this in his right hand, while in his left is a thyrsus, which is a stick of fennel wrapped in ivy and vine-leaves, tipped with a pine cone.

Centaurus contains the closest star to the Sun, Alpha Centauri, 4.3 light years away. Alpha Centauri is also known as Rigil Kentaurus, from the Arabic meaning 'centaur's foot'; Ptolemy described it as lying on the end of the right front leg. To the naked eye it appears as the third-brightest star in the sky, outshone only by Sirius and Canopus, but a small telescope reveals it to be double, consisting of two yellow stars like the Sun. A third, much fainter companion star is called Proxima Centauri because it is just over a tenth of a light year closer to us than the other two, but is only visible telescopically.

Beta Centauri is called Hadar, from an Arabic name signifying one member of a pair of stars; Ptolemy described it as lying on the centaur's left knee. Alpha and Beta Centauri act as pointers to Crux, the Southern Cross, which lies under the centaur's rear quarters. In Ptolemy's day, the stars of Crux were part of Centaurus, as shown on Bayer's illustration on the facing page.

Centaurus also contains the largest and brightest globular star cluster visible from Earth, Omega Centauri. Ptolemy, in the *Almagest*, catalogued this as a 5th-magnitude star on the centaur's back but did not mention its slightly nebulous appearance. Its first identification as a cluster was due to Edmond Halley in 1678. In all, Ptolemy catalogued 37 stars in Centaurus, five of which later became part of Crux.

The changing visibility of Centaurus

It might seem puzzling that Alpha and Beta Centauri and the stars of Crux were known to the ancient Greeks, when they are now too far south to rise above the horizon from Mediterranean latitudes. The reason is the effect known as precession, caused by a wobble of the Earth's axis in space, which slowly changes the position of the celestial poles. In Ptolemy's day, the south celestial pole lay some 10° from where it is now, in a direction away from Centaurus. As a result, the stars of Centaurus and its neighbours were about 10° higher in the Greek sky than they are today. This difference was enough to make them visible from ancient Greece.

Cepheus

Genitive: Cephei
Abbreviation: Cep
Size ranking: 27th
Origin: One of the 48 Greek constellations listed by Ptolemy in the *Almagest*
Greek name: Κηφεύς (Kepheus)

Cepheus was the mythological king of Ethiopia. He was deemed worthy of a place in the sky because he was fourth in descent from the nymph Io, one of the loves of Zeus – and having Zeus as a relative was always an advantage when it came to being commemorated among the constellations. The kingdom of Cepheus was not the Ethiopia we know today, but stretched from the south-eastern shore of the Mediterranean southwards to the Red Sea, an area that contains parts of the modern Israel, Jordan, and Egypt. Ptolemy described him as wearing the tiara-like head-dress of a Persian king.

Cepheus (Κηφεύς in Greek) was married to Cassiopeia, an unbearably vain woman whose boastfulness caused Poseidon to send a sea monster, Cetus, to

Cepheus in the robes and crown of a Persian king, depicted in the *Atlas Coelestis* of John Flamsteed (1729). (University of Michigan Library)

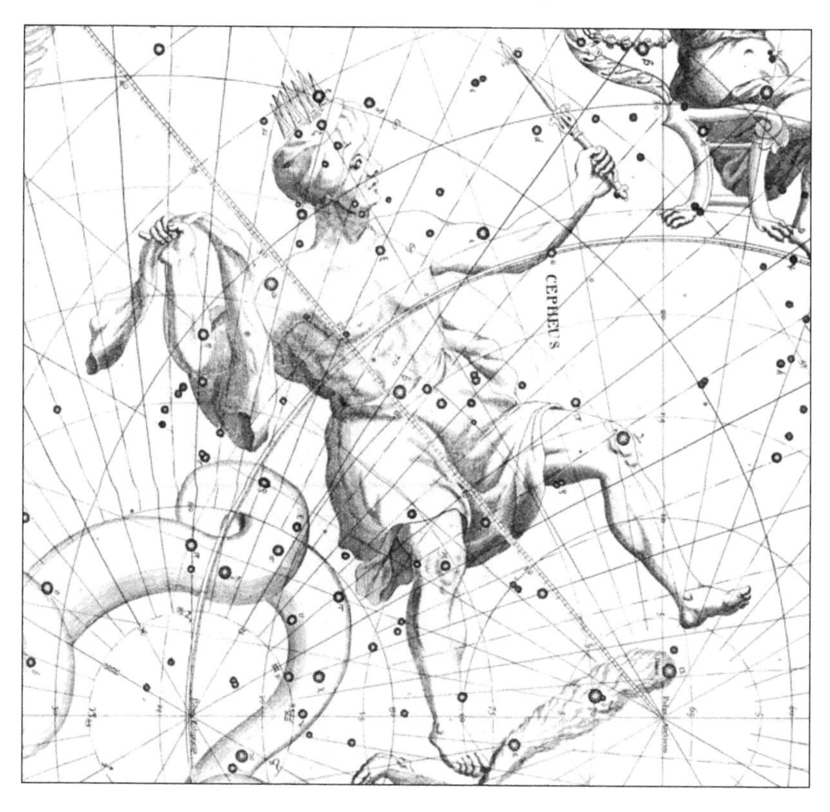

ravage the shores of Cepheus's kingdom. Cepheus was instructed by the Oracle of Ammon to chain his daughter Andromeda to a rock in sacrifice to the monster. She was famously saved from certain death by the hero Perseus, who killed the monster and claimed Andromeda for his bride.

King Cepheus laid on a sumptuous banquet at his palace to celebrate the wedding. But Andromeda had already been promised to Phineus, brother of Cepheus. While the celebrations were in progress, Phineus and his followers burst in, demanding that Andromeda be handed over, which Cepheus refused to do. The dreadful battle that ensued is described in gory detail by Ovid in Book V of his *Metamorphoses*. Cepheus retired from the scene, muttering that he had done his best, and left Perseus to defend himself. Perseus cut down many of his attackers, turning the remainder to stone by showing them the Gorgon's head.

In the sky, the long-suffering Cepheus stands next to Cassiopeia, his feet extending almost to the north celestial pole. Each night he circles the pole and as seen from Greek latitudes would have plunged head first into the sea at the lowest point, suffering the same unceremonious dunking as his vainglorious wife.

The constellation's brightest star is Alpha Cephei, named Alderamin, magnitude 2.5. But its most celebrated star is Delta Cephei, a pulsating supergiant that varies in brightness every 5.4 days; it is the prototype of the Cepheid variable stars that astronomers use for estimating distances in space.

Cetus

The sea monster

Genitive: Ceti
Abbreviation: Cet
Size ranking: 4th
Origin: One of the 48 Greek constellations listed by Ptolemy in the *Almagest*
Greek name: Κῆτος (Ketos)

When Cassiopeia, wife of King Cepheus of Ethiopia, boasted that she was more beautiful than the sea nymphs called the Nereids she set in motion one of the most celebrated stories in mythology, the main characters of which are commemorated among the constellations. In retribution for the insult to the Nereids, the sea god Poseidon sent a fearsome monster to ravage the coast of Cepheus's territory. That monster, a dragon of the sea, is represented by the constellation Cetus. To rid himself of the monster, Cepheus was instructed by the Oracle of Ammon to offer up his daughter Andromeda as a sacrifice to the monster. Andromeda was chained to the cliffs at Joppa (the modern Tel-Aviv) to await her terrible fate.

Cetus was visualized by the Greeks as a hybrid creature, with enormous gaping jaws and the forefeet of a land animal, attached to a scaly body with huge coils like a sea serpent. Hence Cetus is drawn on star maps as a most

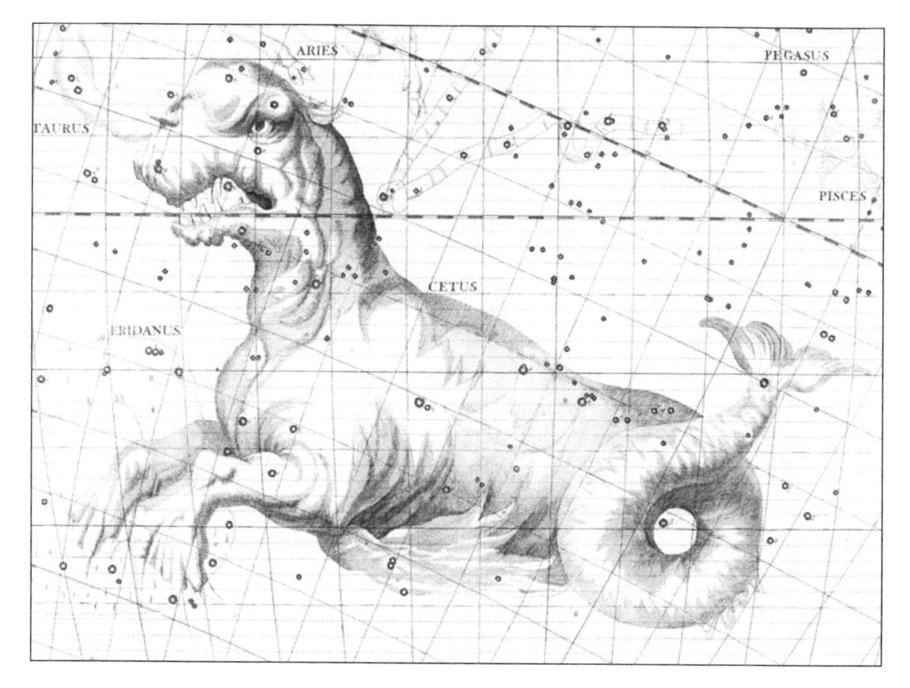

The bizarre-looking sea monster Cetus emerging from the ocean of the southern skies, illustrated in the *Atlas Coelestis* of John Flamsteed (1729). (University of Michigan Library)

unlikely looking creature, more comical than frightening, nothing like a whale although it is sometimes identified as one.

Andromeda trembled as the B-movie monster made towards her, cleaving through the waves like a huge ship. Fortunately, at this moment the hero Perseus happened by and sized up the situation. Swooping down like an eagle onto the creature's back, Perseus drove his diamond-hard sword deep into its right shoulder. Agonized and enraged, the wounded monster reared up on its coils and twisted around, its cruel jaws snapping at its attacker. Again and again Perseus plunged his sword into the beast – through its ribs, its barnacle-encrusted back and at the root of its tail. Spouting blood, the monster finally collapsed into the sea and lay there like a waterlogged hulk. Its corpse was hauled on shore by the appreciative locals who skinned it and put its bones on display.

Cetus (Κῆτος, i.e. Ketos, in Greek) is the fourth-largest constellation, as befits such a monster, but it is not particularly prominent. Its brightest star is second-magnitude Beta Ceti, named Diphda. Ptolemy in the *Almagest* described this star as lying on the end of the southern tail fin; the northern fin was marked by the star we now know as Iota Ceti. Alpha Ceti is called Menkar from the Arabic meaning 'nostrils', a misnomer since this star lies on the beast's jaw – in Ptolemy's description, the star on the nostrils was actually the one to the north, now known as Lambda Ceti.

The most celebrated star in the constellation is Mira, a Latin name meaning 'the amazing one', given on account of its variability in brightness. At times it can easily be seen with the naked eye, but for most of the time it is so faint that it cannot be seen without binoculars or a telescope. Mira is a red giant star whose brightness variations are caused by changes in size. The star was first recorded in 1596 by the Dutch astronomer David Fabricius, but the cyclic nature of the changes was not recognized until 1638. The name Mira was given to the star by the Polish astronomer Johannes Hevelius in 1662, when it was the only variable star known.

Chamaeleon

The chameleon

Genitive: Chamaeleontis
Abbreviation: Cha
Size ranking: 79th
Origin: The 12 southern constellations of Keyser and de Houtman

The celestial chameleon, named after the lizard that can change its skin colour to match its mood, is one of the constellations representing exotic animals introduced by the Dutch navigators Pieter Dirkszoon Keyser and Frederick de Houtman when they charted the southern skies in 1595–97. These new southern constellations were first shown on a globe by their fellow Dutchman Petrus Plancius in 1598 and were rapidly adopted by other map makers such as

Chamaeleon as depicted on Chart XX of the *Uranographia* of Johann Bode (1801). Unlike in some other representations it is ignoring the fly, Musca, which lies above its head, off the top of this illustration. (Deutsches Museum)

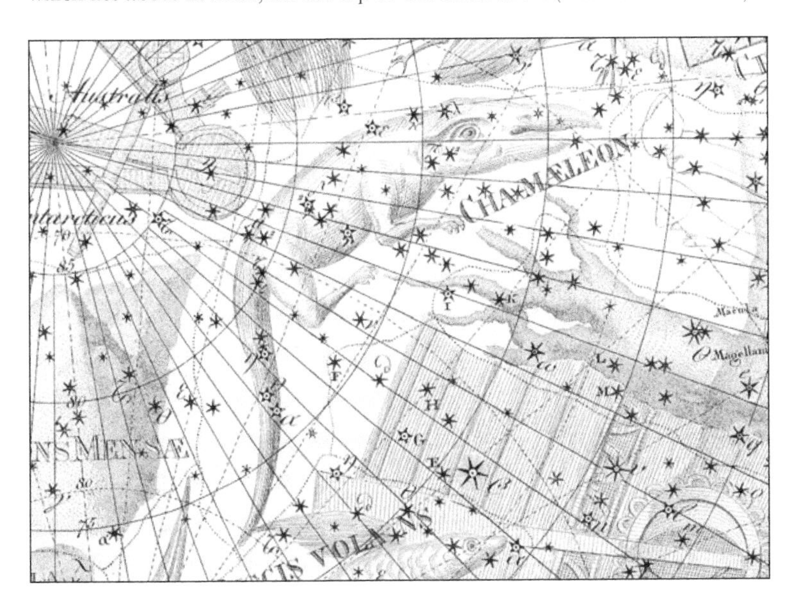

Johann Bayer, for no other observations of the far southern skies were then available. Chameleons are particularly common in Madagascar, where the Dutch fleet stopped to rest and resupply in 1595 on its way to the East Indies, so they probably saw plenty of them there.

Chamaeleon lies near the south celestial pole, next to Musca, the fly. On a globe of 1600 the Dutch cartographer Jodocus Hondius (1563–1612) depicted the chameleon sticking out its tongue to catch the fly. Three years later, Johann Bayer in his *Uranometria* showed the chameleon in the same pose yet evidently failed to appreciate what the adjacent insect, then still unnamed, was supposed to be – he depicted it not as a fly but a bee and named it Apis, as did Bode nearly 200 years later. Chamaeleon has no legends associated with it, and it contains no bright stars.

Circinus

The compasses

Genitive: Circini
Abbreviation: Cir
Size ranking: 85th
Origin: The 14 southern constellations of Nicolas Louis de Lacaille

An insignificant constellation representing a pair of dividing compasses as used by geometers, draughtsmen, and navigators for drawing circles and measuring distances; they are also known as dividers. Circinus was introduced in the 1750s by the Frenchman Nicolas Louis de Lacaille, who fitted various figures into gaps between the existing constellations of the southern skies. In this case the gap seems to have been almost non-existent, and the compasses are squeezed in their folded position between the forefeet of Centaurus and Triangulum Australe. It

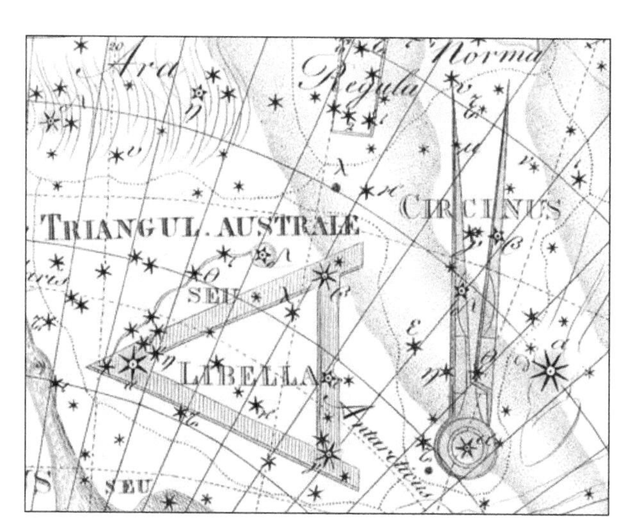

Circinus from Chart XX of the *Uranographia* of Johann Bode, with Triangulum Australe to its left and Norma et Regula (the set square and ruler) above, just off the top of the picture. The bright star to the right of Circinus is Alpha Centauri. (Deutsches Museum)

is the smallest of Lacaille's 14 inventions, and the fourth-smallest constellation in the entire sky.

The constellation first appeared under the French name le Compas on Lacaille's preliminary chart of the southern skies published by the Académie Royal des Sciences in 1756. Its name was Latinized to Circinus on his second chart of 1763. The instrument is conveniently placed next to Triangulum Australe (a pre-existing constellation formed by Keyser and de Houtman which Lacaille visualized as a surveyor's level), and Norma the set square, another of Lacaille's inventions, thereby forming a related set of instruments.

Columba

The dove

Genitive: Columbae
Abbreviation: Col
Size ranking: 54th
Origin: Petrus Plancius

A constellation formed in the late 16th century by the Dutch cartographer and astronomer Petrus Plancius, who took some stars that Ptolemy had catalogued in the *Almagest* as lying outside Canis Major. These unformed stars can be seen, for example, on the southern half of Albrecht Dürer's star chart (page 11) as

Columba flies with an olive branch in its beak as seen on Chart XVIII of the *Uranographia* of Johann Bode (1801). (Deutsches Museum)

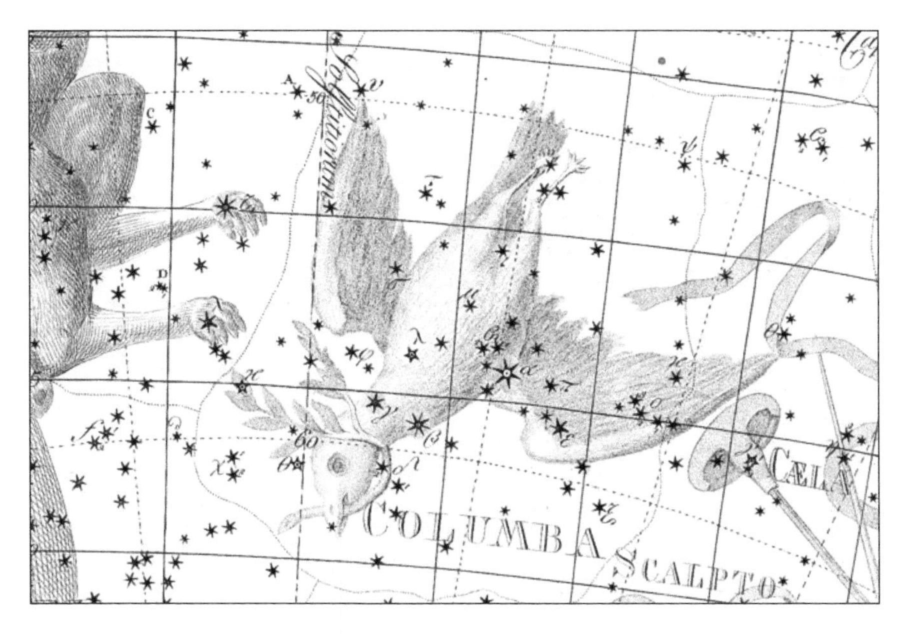

two little groups, one to the south of Lepus and the other between the hind legs of Canis Major. Columba was first shown as a separate constellation by Plancius in 1592 on a celestial hemisphere that he tucked into the corner of his first great terrestrial map, although it was unnamed. It flies behind Argo Navis, the ship, and under the hind legs of Canis Major.

Columba is supposed to represent Noah's dove, sent out from the Ark to find dry land, and which returned with an olive branch in its beak, a sign that the Flood was at last subsiding. To complete this Biblical tableau, Plancius even renamed Argo as Noah's Ark on a globe of 1613. However, those familiar with the story of Argo (page 187) might instead think of Columba as the dove sent by the Argonauts between the sliding doors of the Clashing Rocks to ensure their safe passage. The constellation's brightest star, third-magnitude Alpha Columbae, is called Phact, from an Arabic name meaning 'ring dove'.

Coma Berenices

Berenice's hair

Genitive: Comae Berenices
Abbreviation: Com
Size ranking: 42nd
Origin: Caspar Vopel

Between Boötes and Leo lies a fan-shaped swarm of faint stars that was known to the Greeks but was not classed by them as a separate constellation, being considered part of Leo. Eratosthenes referred to the swarm as the hair of Ariadne under his entry on the Northern Crown (Corona Borealis), but under Leo he said it was the hair of Queen Berenice of Egypt, which is how we identify it today. Ptolemy, in his *Almagest* entry on Leo, referred to the swarm as 'a nebulous mass, called the lock' (i.e. of hair). He listed three stars at the corners of the swarm. This triangle of stars can be seen, for example, on Albrecht Dürer's star chart of 1515, above the tail of Leo and behind the rear legs of the Great Bear (see page 10).

The group was first shown as a separate constellation in 1536, under the name Berenices Crinis, on a globe by the German mathematician and cartographer Caspar Vopel (1511–61). He was followed in 1551 by the Dutch cartographer Gerardus Mercator (1512–94) who termed the constellation Cincinnus, a Latin word meaning lock of hair. In 1602 Tycho Brahe included Coma Berenices in his influential star catalogue, thus ensuring its widespread adoption. The constellation as envisaged by Vopel, Mercator, and Tycho covers a much larger area than the nebulous mass described by Ptolemy, which is now known as the cluster Melotte 111.

The modern constellation's three brightest stars, all of fourth magnitude, were labelled Alpha, Beta, and Gamma by Francis Baily in his *British Association Catalogue* of 1845.

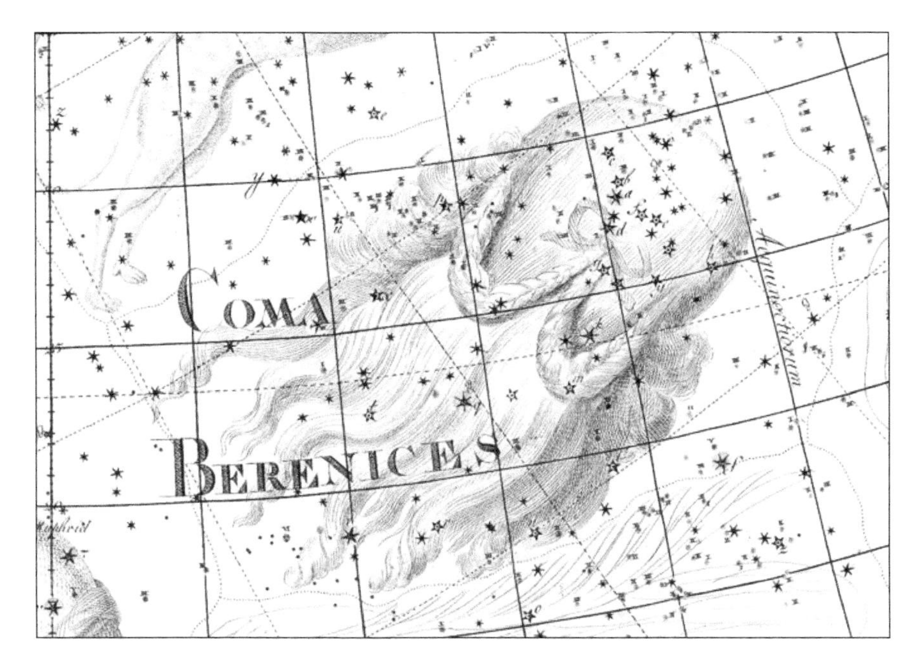

Coma Berenices, the flowing tresses of an Egyptian queen, from Chart VII of Johann Bode's *Uranographia* (1801). The 'nebulous mass' described by Ptolemy that formed the basis of the constellation (actually a scattering of faint stars) is in the upper right part of the illustration, on the crown of the hair. (Deutsches Museum)

A hair-raising story

Berenice was a real person who, in 246 BC, married her cousin, Ptolemy III Euergetes (Hyginus says she was his sister, but that was a different Berenice). Berenice was reputedly a great horsewoman who had already distinguished herself in battle. Hyginus, who deals with the star group under Leo in his *Poetic Astronomy*, tells the following story.

It seems that shortly after their marriage (Hyginus says a few days, although actually it was a few months) Ptolemy set out to attack Asia on the Third Syrian War. Berenice vowed that if he returned victorious she would cut off her hair in gratitude to the gods. On Ptolemy's safe return the following year, the relieved Berenice carried out her promise and placed her hair in the temple dedicated to her mother Arsinoë (identified after her death with Aphrodite) at Zephyrium near the modern Aswan. But the following day the tresses were missing. What really happened to them is not recorded, but Conon of Samos (*c*.280–*c*.220 BC), a mathematician and astronomer who worked at Alexandria, pointed out the group of stars near the tail of the lion, telling the King that the hair of Berenice had gone to join the constellations.

In reality, the disappearance of the hair and its subsequent 'discovery' among the stars was most likely staged to glorify Ptolemy and his queen among their subjects. The story was mythologized by the court poet Callimachus (*c*.305–*c*.240 BC) in his popular poem called *Lock of Berenice*.

Corona Australis

The southern crown

Genitive: Coronae Australis
Abbreviation: CrA
Size ranking: 80th
Origin: One of the 48 Greek constellations listed by Ptolemy in the *Almagest*
Greek name: Στέφανος νότιος (Stephanos notios)

Corona Australis was known to the Greeks not as a crown but as a wreath, which is how it is depicted on old star maps. Aratus did not name it as a separate constellation but referred to it simply as a circlet of stars beneath the forefeet of Sagittarius. Perhaps it has slipped off the archer's head.

The first recorded mention of Corona Australis as a separate constellation appears to be in the first century BC by the Greek astronomer and mathematician Geminus of Rhodes in his survey of astronomy called *Introduction to the Phenomena*. He gave it the name Νότιος στέφανος (Notios stephanos, the southern crown), whereas Ptolemy in the *Almagest* two centuries later reversed the name as Στέφανος νότιος (Stephanos notios). The northern crown, which we call Corona Borealis, was known to the Greeks simply as Στέφανος (Stephanos).

None of the stars of Corona Australis is brighter than fourth magnitude and there seem to be no legends associated with it, unless this is the crown placed in the sky by Dionysus after retrieving his dead mother from the Underworld. Hyginus gives this myth under the Northern Crown (Corona Borealis) but it seems out of place there and he may have confused the two constellations. If so, the wreath would be made of myrtle leaves, for Dionysus left a gift of myrtle in Hades in return for his mother, and the followers of Dionysus wore crowns of myrtle.

Corona Australis lies at the forefeet of Sagittarius, the archer, as depicted on Chart XV of the *Uranographia* of Johann Bode (1801). (Deutsches Museum)

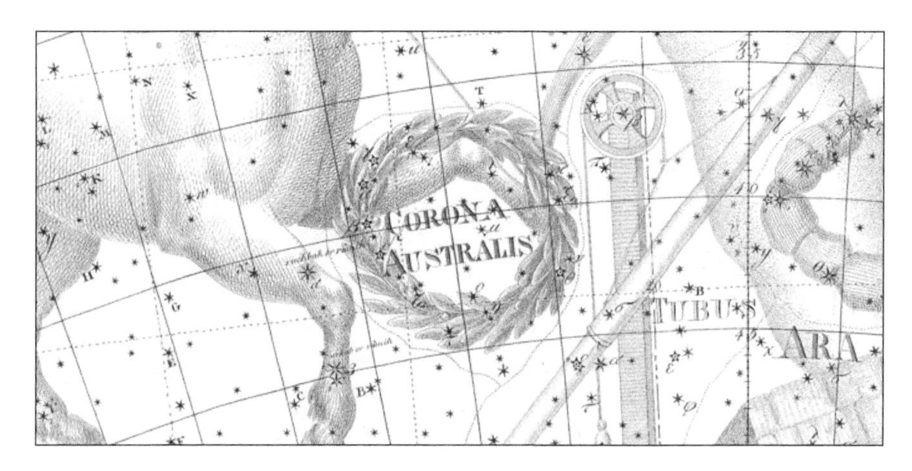

Corona Borealis

The northern crown

Genitive: Coronae Borealis
Abbreviation: CrB
Size ranking: 73rd
Origin: One of the 48 Greek constellations listed by Ptolemy in the *Almagest*
Greek name: Στέφανος (Stephanos)

A semicircle of stars between Boötes and Hercules marks the golden crown worn by Princess Ariadne of Crete when she married the god Dionysus. The crown is said to have been made by Hephaestus, the god of fire, and was studded with jewels from India.

Ariadne, daughter of King Minos of Crete, is famous in mythology for her part in helping Theseus to slay the Minotaur, the gruesome creature with the head of a bull on a human body. Ariadne was actually half-sister to the Minotaur, for her mother Pasiphae had given birth to the creature after copulating with a bull owned by King Minos. To hide the family's shame, Minos imprisoned the Minotaur in a labyrinth designed by the master craftsman Daedalus. So complex was the maze of the labyrinth that neither the Minotaur nor anyone else who ventured in could ever find their way out.

One day the hero Theseus, son of King Aegeus of Athens, came to Crete. Theseus was a strong, handsome man with many of the qualities of Heracles

Corona Borealis, the jewelled crown of Princess Ariadne, shown in the *Atlas Coelestis* of John Flamsteed (1729). (University of Michigan Library)

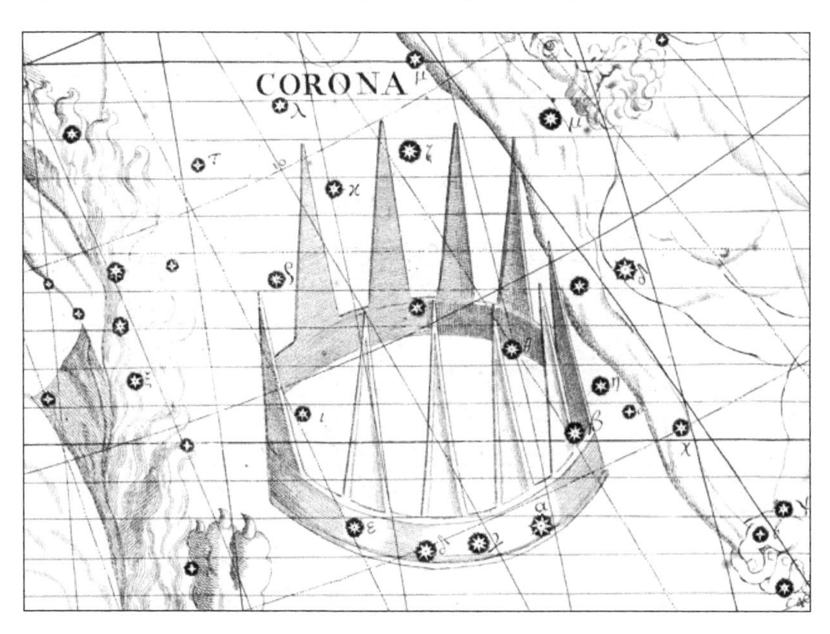

and was unsurpassed as a wrestler. Ariadne fell in love with him on sight. When Theseus offered to kill the Minotaur she consulted Daedalus, who gave her a ball of thread and advised Theseus to tie one end to the door of the labyrinth and pay out the thread as he went along. After killing the Minotaur with his bare hands, Theseus emerged by following the trail of thread back to the door.

He sailed off with Ariadne, but no sooner had they reached the island of Naxos than he abandoned her. As she sat there, cursing Theseus for his ingratitude, she was seen by Dionysus. The god's heart melted at the sight of the forlorn girl and he married her on the spot.

Accounts differ about where Ariadne's crown came from. One story says that it was given to her by Aphrodite as a wedding present. Others say that Theseus obtained it from the sea nymph Thetis, and that its sparkling light helped Theseus find his way through the labyrinth. Whatever the case, after their wedding Dionysus joyfully tossed the crown into the sky where its jewels transformed into stars.

The Greeks knew Corona as Στέφανος (Stephanos), meaning 'crown' or 'wreath'. Its brightest star, second-magnitude Alpha, is officially called Alphecca from the Arabic name for the constellation, although it was once also known as Gemma, the Latin for 'jewel'.

Corvus
The crow

and

Crater
The cup

Genitives: Corvi; Crateris
Abbreviations: Crv; Crt
Size rankings: 70th; 53rd
Origin: Two of the 48 Greek constellations listed by Ptolemy in the *Almagest*
Greek names: Κόραξ (Korax); Κρατήρ (Krater)

These two adjacent constellations are linked in a moral tale that goes back at least to the time of Eratosthenes in the third century BC. As told by Ovid in his *Fasti*, Apollo was about to make a sacrifice to Zeus and sent the crow to fetch water from a running spring. The crow flew off with a bowl in its claws until it came to a fig tree laden with unripe fruit. Ignoring its orders, the crow waited several days for the fruit to ripen, by which time Apollo had been forced to find a source of water for himself.

After eating its fill of the delicious fruit, the crow looked around for an alibi. He picked up a water-snake in his claws and returned with it to Apollo, blaming the serpent for blocking the spring. But Apollo, one of whose skills was the art

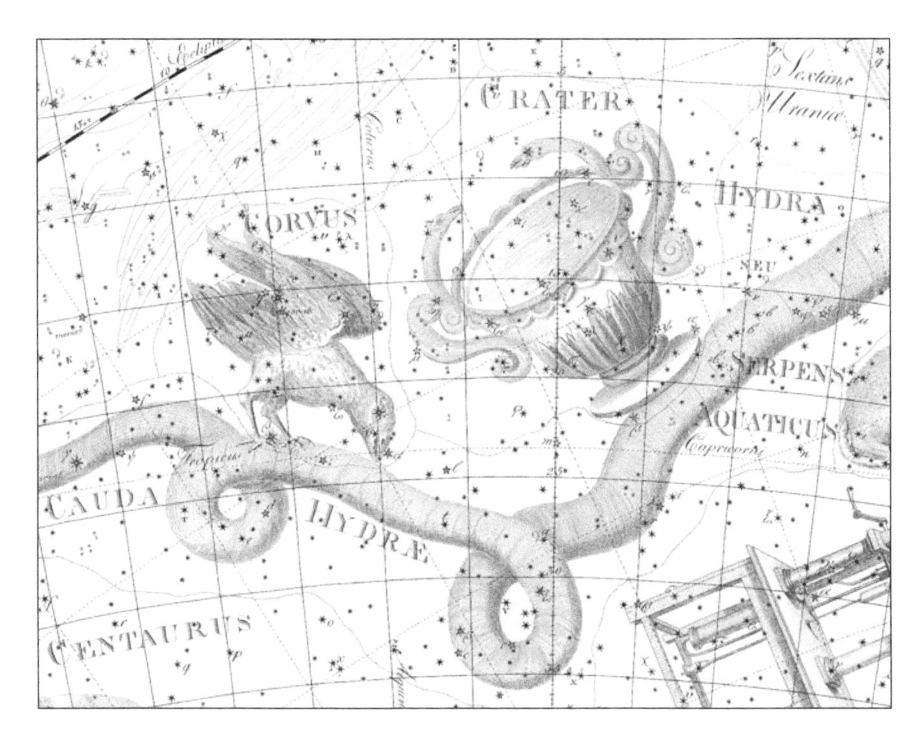

Corvus and Crater, adjacent constellations on the back of Hydra, seen on Chart XIX of the *Uranographia* of Johann Bode (1801). In ancient Greece, a krater was a bowl or vase used for mixing wine with water, rather than a cup as we know it. (Deutsches Museum)

of prophecy, saw through the lie and condemned the crow to a life of thirst – which is perhaps one explanation for the rasping call of the crow. In memorial of this incident Apollo put the crow, the cup, and the water-snake together in the sky.

The crow is depicted pecking at the water snake's coils, as though attempting to move it so that the crow may reach the cup to drink. The cup, usually represented as a magnificent double-handed chalice of the type known in Greece as a krater, is shown tilted towards the crow but tantalizingly just out of the thirsty bird's reach. The water-snake is the constellation Hydra which, in another legend, doubles as the creature slain by Heracles.

The crow was the sacred bird of Apollo, who changed himself into one to flee from the monster Typhon when that immense creature threatened the gods. In another story, related by Ovid in his *Metamorphoses*, the crow was once snow-white like a dove, but the bird brought news to Apollo that his love, Coronis, had been unfaithful. Apollo in his anger cursed the crow, turning it forever black.

In Greek, Crater was known as Κρατήρ (Krater) while Corvus was Κόραξ, i.e. Korax. The star we know as Alpha Corvi, named Alchiba, represents the crow's beak, pecking at the Hydra, while Beta marks its feet, planted firmly on the Hydra's back.

Crux

The southern cross

Genitive: Crucis
Abbreviation: Cru
Size ranking: 88th
Origin: Petrus Plancius

The smallest of all the 88 constellations. The stars of Crux were known to the ancient Greeks, and were catalogued by Ptolemy in the *Almagest*, but were regarded as part of the hind legs of Centaurus, the centaur, rather than as a separate constellation. They subsequently became lost from view to Europeans because of the effect of precession, which causes a gradual drift in the position of the celestial pole against the stars, and were rediscovered during the 16th century by seafarers venturing south.

The Italian explorer Amerigo Vespucci (1454–1512) charted what seems to have been Alpha and Beta Centauri and the stars of Crux in 1501, but the most accurate early depiction was made by the Italian navigator Andrea Corsali (1487–??) in 1515. Corsali described the pattern as 'so fair and beautiful that no other heavenly sign may be compared to it'. Thereafter navigators began using the cross for direction finding, since its long axis from Gamma via Alpha Crucis

Crux lies under the hind legs of Centaurus in a rich area of the Milky Way. It contains a dark cloud of dust, here shown white, known to modern astronomers as the Coalsack but named Macula Magellanica on this illustration from Chart XX of the *Uranographia* star atlas of Johann Bode (1801). (Deutsches Museum)

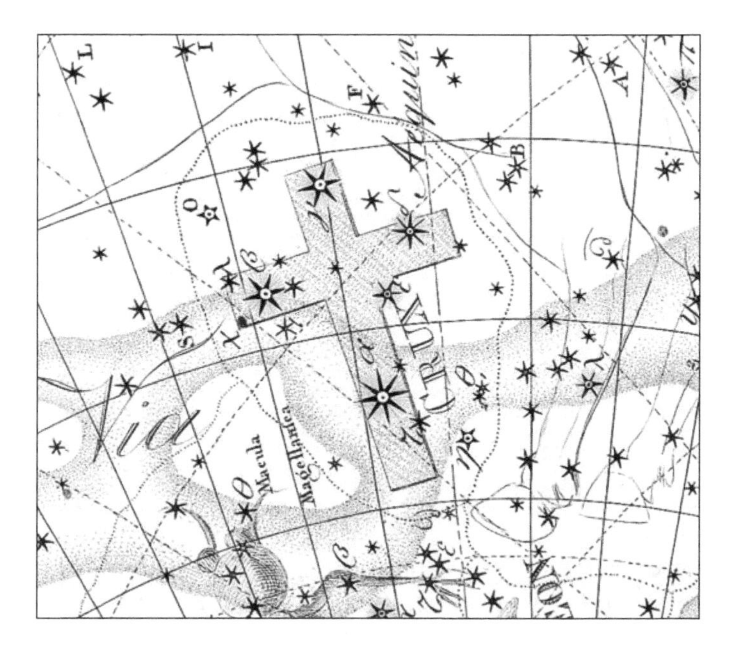

points to the south celestial pole, and it was adopted by astronomers as a separate constellation by the end of the 16th century.

Crux first appeared in its modern form on the celestial globes by the Dutch cartographers Petrus Plancius and Jodocus Hondius in 1598 and 1600; Plancius had earlier shown a stylized southern cross in a completely different part of the sky, south of Eridanus. It seems that only after he received the first accurate observations of the southern stars made by the Dutch navigator Pieter Dirkszoon Keyser did Plancius realize that the stars of Crux had been listed by Ptolemy all along, as part of Centaurus. Benefiting from this revelation, Johann Bayer drew the cross over the hind legs of Centaurus on his *Uranometria* atlas of 1603 (see illustration on page 70).

The five main stars of Crux were listed as a separate constellation for the first time under the name De Cruzero in the southern star catalogue of another Dutch seafarer, Frederick de Houtman, published in 1603 (see box page 20). The first printed chart to show Crux separately from Centaurus was that of the German astronomer Jacob Bartsch (*c.*1600–33), brother-in-law of Johannes Kepler, in his book *Usus Astronomicus Planisphaerii Stellati* (Astronomical Use of the Stellar Planisphere) published in 1624.

The constellation's brightest star is called Acrux, a name originally applied by navigators from its scientific designation Alpha Crucis. At declination −63.1°, Acrux is the most southerly first-magnitude star. The names Becrux and Gacrux for Beta and Gamma Crucis have a similar modern origin, although Becrux is now only an unofficial alternative to its IAU-recognized name of Mimosa.

Crux contains a famous dark cloud of gas and dust called the Coalsack Nebula, which appears in silhouette against the bright Milky Way background. This was first described in an account by Amerigo Vespucci published in 1503 or 1504, where it was described as a 'black canopus of immense bigness'. It lies some 600 light years away and spills over the borders of Crux into adjoining Centaurus and Musca.

Cygnus

The swan

Genitive: Cygni
Abbreviation: Cyg
Size ranking: 16th
Origin: One of the 48 Greek constellations listed by Ptolemy in the *Almagest*
Greek name: Ὄρνις (Ornis)

A popular name for Cygnus is the Northern Cross, and indeed its shape is far larger and more distinctive than the famous Southern Cross. In its cruciform shape the Greeks visualized the long neck, outstretched wings and stubby tail of a swan flying along the Milky Way, in which it is embedded. Aratus's description of it as being hazy or misty in parts is no doubt a reference to the Milky

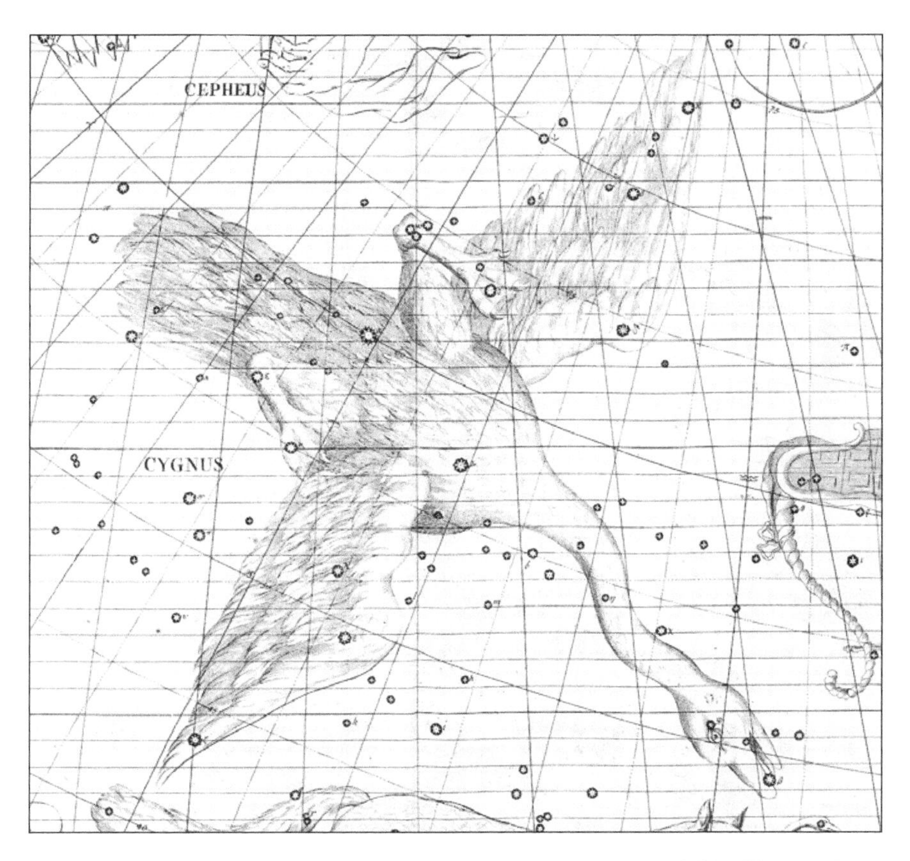

Cygnus flying down the Milky Way in a chart from the *Atlas Coelestis* of John Flamsteed (1729). At the root of its tail lies the bright star Deneb, here labelled simply with the Greek letter Alpha. (University of Michigan Library)

Way running through it. The mythographers tell us that the swan is Zeus in disguise, on his way to one of his innumerable love affairs, but his exact quarry is a subject of some disagreement.

The version of the tale that goes back to Eratosthenes in the third century BC says that Zeus one day took a fancy to the nymph Nemesis, who lived at Rhamnus, some way north-east of Athens. To escape his unwelcome advances she assumed the form of various animals, first jumping into a river, then fleeing across land, before finally taking flight as a goose. Not to be outdone, Zeus pursued her through all these transformations, at each step turning himself into a larger and swifter animal, until he finally became a swan in which form he caught and raped her.

Hyginus tells a similar story, but does not mention the metamorphoses of Nemesis. Rather, he says that Zeus pretended to be a swan escaping from an eagle and that Nemesis gave the swan sanctuary. Only after she had gone to sleep with the swan in her lap did she discover her mistake. In both versions the outcome was that Nemesis produced an egg which was then given to Queen

Leda of Sparta, some say by Hermes and others say by a passing shepherd who found the egg in a wood. Out of the egg hatched the beautiful Helen, later to become famous as Helen of Troy.

Leda and the swan

A simpler alternative says that Zeus seduced Leda in the form of a swan by the banks of the river Eurotas; with this story in mind, Germanicus Caesar refers to the swan as the 'winged adulterer'. Leda was the wife of King Tyndareus of Sparta, which considerably complicated the outcome because she also slept with her husband later the same night.

According to one interpretation, she gave birth to a single egg from which hatched the twins Castor and Polydeuces as well as Helen. The shell of this egg was said to have been put on display at a temple in Sparta, hanging by ribbons from the ceiling. A rival account says that Leda produced two eggs, from one of which emerged Castor and Polydeuces while from the other came Helen and her sister Clytemnestra. To add to the confusion, Polydeuces and Helen were reputedly the children of Zeus, while Castor and Clytemnestra were fathered by Tyndareus. Castor and Polydeuces are commemorated by the constellation Gemini, where Polydeuces is better known by his Latin name, Pollux.

Aratus called the constellation simply Ὄρνις (Ornis), the bird. Its identification with a swan (Κύκνος, i.e. Kyknos, or Cygnus in Latin transliteration) was introduced half a century later by Eratosthenes. Ptolemy, however, in his *Almagest* ignored Eratosthenes's identification of the constellation as a swan and instead followed Aratus by referring to it as an anonymous Bird.

Deneb and Albireo – plus a black hole

Cygnus's brightest star, Deneb, marks the tail of the swan; its name comes from *dhanab*, the Arabic word for 'tail'. Surprisingly, the Greeks had no name for this prominent star. Deneb is a highly luminous supergiant about 1,400 light years away, the most distant of all first-magnitude stars. It forms one corner of the so-called Summer Triangle of stars completed by Vega in the constellation Lyra and Altair in Aquila.

The beak of the swan is marked by a star named Albireo, revealed by small telescopes to be a beautiful coloured double star of green and amber, like a celestial traffic light. The German historian Paul Kunitzsch has traced the tortuous history of the name Albireo. It started with an Arabic translation of the Greek word for 'bird', Ὄρνις (ornis), the name which Aratus and Ptolemy gave for the constellation. In the Middle Ages this Arabic name was mistranslated back into Latin, where it was described as 'ab ireo', meaning that it was thought to come from the name of a certain herb. This phrase was mistaken for an Arabic name and was rewritten as *albireo*. Hence although the name Albireo looks Arabic, it is completely meaningless.

Cygnus lies in the Milky Way and contains many attractive star fields for sweeping with binoculars. Its most celebrated object cannot be seen by optical means at all: a black hole 6,000 light years away called Cygnus X-1 which lies halfway along the swan's neck.

Delphinus
The dolphin

Genitive: Delphini
Abbreviation: Del
Size ranking: 69th
Origin: One of the 48 Greek constellations listed by Ptolemy in the *Almagest*
Greek name: Δελφίν (Delphin)

Dolphins were a familiar sight to Greek sailors, so it is not surprising to find one of these friendly and intelligent creatures depicted in the sky. Two stories account for the presence of the celestial dolphin. According to Eratosthenes, this jaunty dolphin represents the messenger of the sea god Poseidon.

After Zeus, Poseidon, and Hades had overthrown their father Cronus, they divided up the sky, the sea, and the underworld between them, with Poseidon inheriting the sea. He built himself a magnificent underwater palace off the island of Euboea. For all its opulence, the palace felt empty without a wife, so Poseidon set out in search of one. He courted Amphitrite, one of the group of sea nymphs called Nereids, but she fled from his rough advances and took refuge among the other Nereids. Poseidon sent messengers after her, including a dolphin, which found her and with soothing gestures brought her back to the

A playful-looking Delphinus leaps among the stars in the *Atlas Coelestis* of John Flamsteed (1729). (University of Michigan Library)

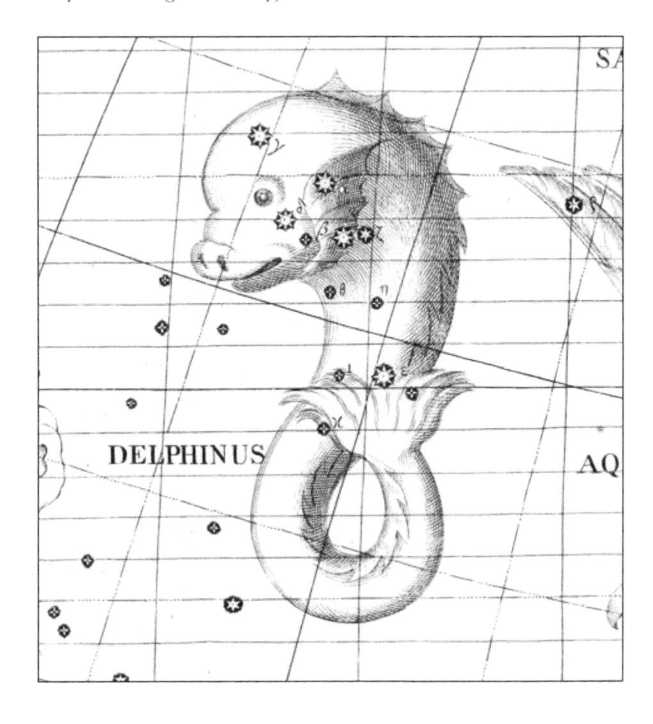

sea god, whom she subsequently married. In gratitude, Poseidon placed the image of the dolphin among the stars.

Another story, given by Hyginus and Ovid, says that this is the dolphin that saved the life of Arion, a real-life poet and musician of the seventh century BC. Arion was born on the island of Lesbos, but his reputation spread throughout Greece for he was said to be unequalled in his skill with the lyre. While Arion was returning to Greece by ship from a concert tour of Sicily and southern Italy, the sailors plotted to kill him and steal the small fortune that he had earned.

When the sailors surrounded him with swords drawn, Arion asked to be allowed to sing one last song. His music attracted a school of dolphins which swam alongside the ship, leaping playfully. Placing his faith in the gods, Arion leaped overboard – and one of the dolphins carried him on its back to Greece, where Arion later confronted his attackers and had them sentenced to death. Apollo, god of music and poetry, placed the dolphin among the constellations, along with the lyre of Arion which is represented by the constellation Lyra. Delphinus was known in Greek as Δελφίν (Delphin) or Δελφίς (Delphis).

Two stars in Delphinus bear the peculiar names Sualocin and Rotanev, which first appeared in the *Palermo Catalogue* of 1814 compiled by the Italian astronomer Giuseppe Piazzi (1746–1826). Read backwards, these names spell out Nicolaus Venator, the Latinized form of Niccolò Cacciatore (1770–1841), who was Piazzi's assistant and eventual successor at Palermo Observatory. It is usually said that Cacciatore was responsible for the naming, which would make him the only person to have named a star after himself and got away with it. However, it is equally possible that the names were applied by Piazzi to honour his heir apparent, or 'dauphin' (dolphin).

The constellation was once popularly called Job's Coffin, presumably from its elongated box-like shape, although sometimes this name is restricted to the diamond formed by the four stars Alpha, Beta, Gamma, and Delta Delphini. Who originated the name Job's Coffin, or when, is not known.

Dorado

The goldfish

Genitive: Doradus
Abbreviation: Dor
Size ranking: 72nd
Origin: The 12 southern constellations of Keyser and de Houtman

A small southern constellation introduced at the end of the 16th century by the Dutch navigators Pieter Dirkszoon Keyser and Frederick de Houtman. Dorado was first depicted on a star globe of 1598 by the Dutchman Petrus Plancius and first appeared in print in 1603 on the *Uranometria* atlas of Johann Bayer.

The constellation represents the colourful dolphinfish *Coryphaena hippurus* (also known as mahi-mahi) found in tropical waters, not the goldfish commonly found

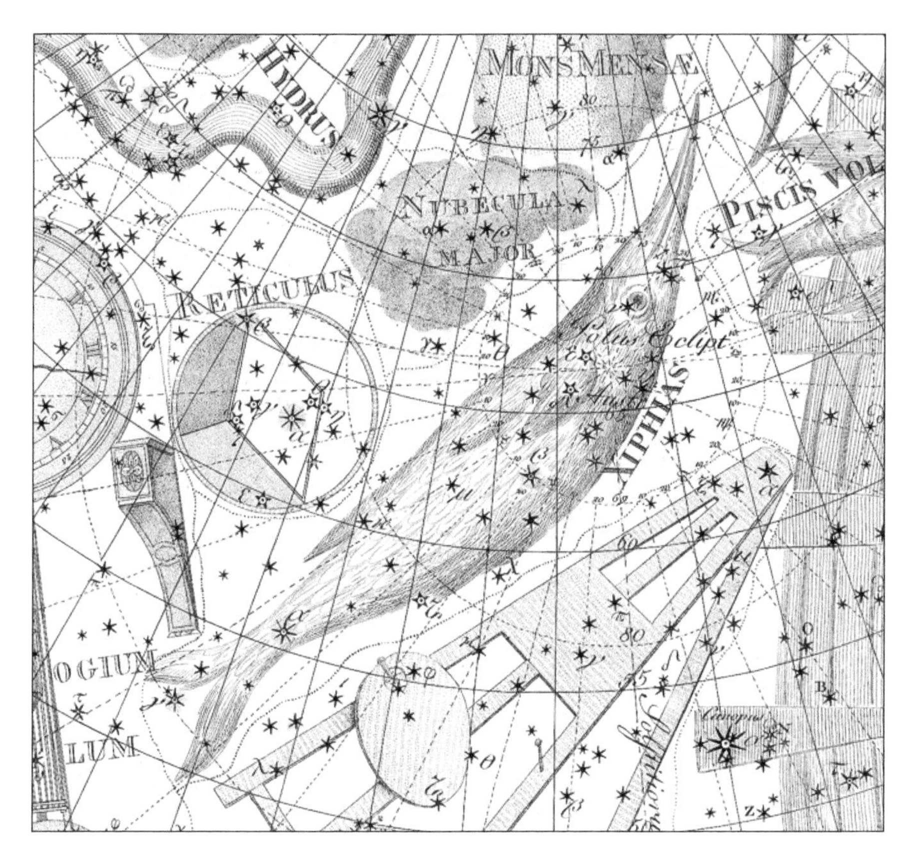

Dorado shown on Chart XX of the *Uranographia* atlas of Johann Bode under the name of Xiphias, the swordfish. Nubecula Major, above it, is better known as the Large Magellanic Cloud. (Deutsches Museum)

in ponds and aquaria. Dutch explorers observed these large predatory fish chasing flying fish and so Dorado was placed in the sky following the constellation of the flying fish, Volans (page 181). The constellation has also been known as Xiphias, the Swordfish, a name that first appeared as an alternative to Dorado in the *Rudolphine Tables* of Johannes Kepler published in 1627. Johann Bode depicted it as Xiphias on his *Uranographia* star atlas of 1801 (see above).

Dorado's main claim to fame is that it contains most of the Large Magellanic Cloud, a small neighbour galaxy of our own Milky Way, about 170,000 light years away; this, like the Small Magellanic Cloud in Tucana, was first described by the Italian explorer Amerigo Vespucci (1454–1512) in an account published in 1503 or 1504. An old name for the Large Magellanic Cloud was Nubecula Major, as used on Bode's atlas.

Within the Large Magellanic Cloud lies the huge nebula NGC 2070, popularly called the Tarantula. It is also known as 30 Doradus or the 30 Doradus Nebula; this is its number in Bode's star catalogue *Allgemeine Beschreibung und Nachweisung der Gestirne*, published in 1801 to accompany his *Uranographia* atlas.

Draco

The dragon

Genitive: Draconis
Abbreviation: Dra
Size ranking: 8th
Origin: One of the 48 Greek constellations listed by Ptolemy in the *Almagest*
Greek name: Δράκων (Drakon)

Coiled around the sky's north pole is the celestial dragon, Draco, known to the Greeks as Δράκων (Drakon). Legend has it that this is the dragon slain by Heracles during one of his labours, and in the sky the dragon is depicted with one foot of Heracles (in the form of the neighbouring constellation Hercules)

Draco winds around the north celestial pole on Chart III of Johann Bode's *Uranographia* star atlas (1801). The dragon's long tail, labelled Cauda Draconis on this chart, extends between the two bears. (Deutsches Museum)

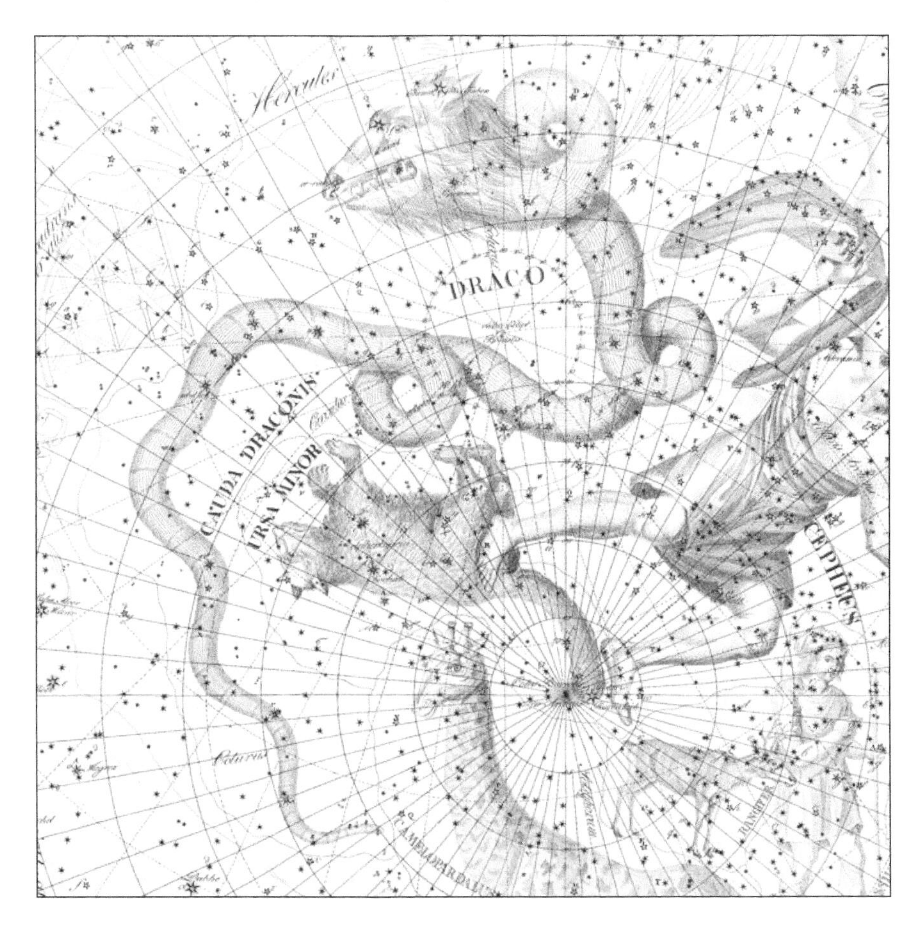

planted firmly upon its head. This dragon, named Ladon, guarded the precious tree on which grew the golden apples.

Hera had been given the golden apple tree as a wedding present when she married Zeus. She was so delighted with it that she planted it in her garden on the slopes of Mount Atlas and set the Hesperides, daughters of Atlas, to guard it. Most authorities say there were three Hesperides, but Apollodorus names four. They proved untrustworthy guards, for they kept picking the apples. Sterner measures were required, so Hera placed the dragon Ladon around the tree to ward off pilferers.

According to Apollodorus, Ladon was the offspring of the monster Typhon and Echidna, a creature half woman and half serpent. Ladon had one hundred heads, says Apollodorus, and could talk in different voices. Hesiod, though, says that the dragon was the offspring of the sea deities Phorcys and Ceto, and he does not mention the number of heads. In the sky, the dragon is single-headed.

The great hero Heracles was required to steal some apples from the tree as one of his labours. He did so by killing the dragon with his poisoned arrows. Apollonius Rhodius recounts that the Argonauts came across the body of Ladon the day after Heracles had shot him. The dragon lay by the trunk of the apple tree, its tail still twitching but the rest of its coiled body bereft of life. Flies died in the poison of its festering wounds while nearby the Hesperides bewailed the dragon's death, covering their golden heads with their white arms. Hera placed the image of the dragon in the sky as the constellation Draco.

Despite its considerable size, the eighth-largest constellation, Draco is not particularly prominent. Its brightest star is second-magnitude Gamma Draconis, called Eltanin from the Arabic *al-tinnin* meaning 'the serpent'. Alpha Draconis is called Thuban, from a highly corrupted form of the Arabic *ra's al-tinnin*, 'the serpent's head'. Beta Draconis is called Rastaban, another corrupted form of the same Arabic name. The stars Beta, Gamma, Nu, and Xi Draconis form a lozenge shape which we regard as the dragon's head, but which bedouin Arabs visualized as four mother camels with a baby camel at the centre, the baby being represented by an unnamed 6th-magnitude star.

Equuleus

The little horse

Genitive: Equulei
Abbreviation: Equ
Size ranking: 87th
Origin: One of the 48 Greek constellations listed by Ptolemy in the *Almagest*
Greek name: Ἵππου Προτομή (Hippou Protome)

This insignificant constellation, second-smallest in the sky, was one of the 48 constellations listed by the Greek astronomer Ptolemy in the second century AD. It was unknown to Aratus 400 years earlier, and its invention is often attributed

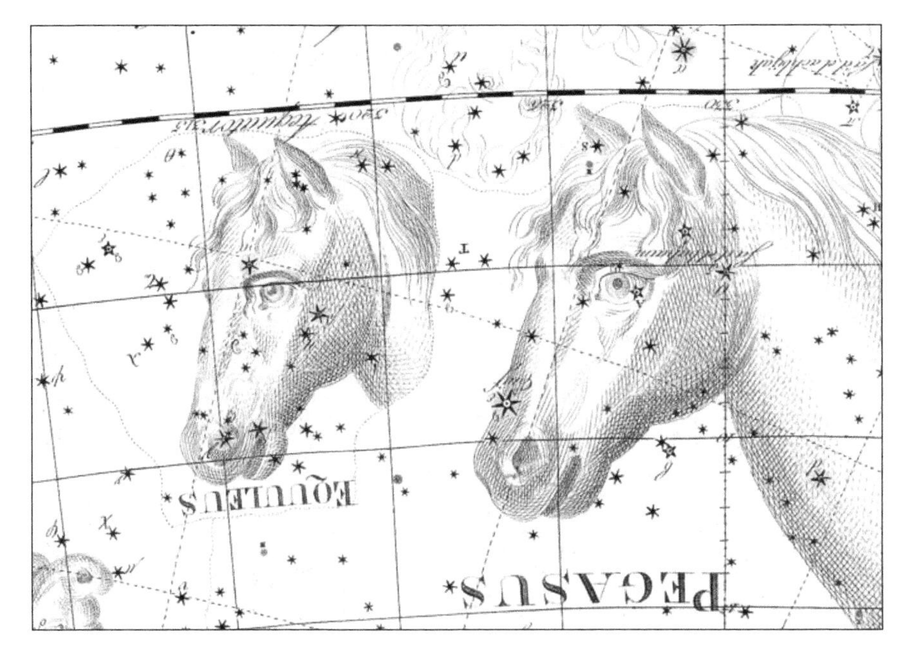

Equuleus, the foal, seen at left next to the head of Pegasus on Chart X of the *Uranographia* of Johann Bode (1801). (Deutsches Museum)

to Ptolemy. In the *Almagest* he called it Ἵππου Προτομή (Hippou Protome), the forepart of a horse; Equuleus is a later Latin name.

However, that is not the whole story. The Greek writer Geminus, who pre-dated Ptolemy by a century or two, tells us that Hipparchus introduced a constellation called Protome hippou in the second century BC. We do not have Hipparchus's original to check, but presumably it was the same constellation that Ptolemy adopted for his own catalogue. Hence the true inventor of Equuleus seems to have been Hipparchus rather than Ptolemy.

Equuleus consists merely of a few stars of fourth magnitude and fainter forming the head of a horse, next to the head of the much better-known horse Pegasus. The early mythologists such as Eratosthenes and Hyginus never mentioned this little horse, but perhaps Ptolemy (or Hipparchus) had in mind the story of Hippe and her daughter Melanippe, sometimes told for Pegasus but which seems more appropriate for Equuleus.

Hippe, daughter of Chiron the centaur, one day was seduced by Aeolus, grandson of Deucalion. To hide the secret of her pregnancy from Chiron she fled into the mountains, where she gave birth to Melanippe. When her father came looking for her, Hippe appealed to the gods who changed her into a mare. Artemis placed the image of Hippe among the stars, where she still hides from Chiron (represented by the constellation Centaurus), with only her head showing.

The fourth-magnitude star Alpha Equulei is called Kitalpha from the Arabic meaning 'the section of the horse', in reference to the constellation as a whole.

Eridanus

The river

Genitive: Eridani
Abbreviation: Eri
Size ranking: 6th
Origin: One of the 48 Greek constellations listed by Ptolemy in the *Almagest*
Greek name: Ποταμός (Potamos)

Aratus applied the mythical name Ἠριδανός (Eridanos) to this constellation although many other authorities, including Ptolemy in the *Almagest*, simply called it Ποταμός (Potamos), meaning river. Eratosthenes had another identification: he said that the constellation represented the Nile, 'the only river which runs from south to north'. Hyginus agreed, claiming that the star Canopus lay at the end of the celestial river, in the same way that the island Canopus lies at the mouth of the Nile. However, in this he was wrong, for Canopus marks a steering oar of the ship Argo and is not part of the river. Hyginus had evidently misunderstood a comment by Eratosthenes, who had simply said that Canopus lay 'beneath' the river, meaning that it was at a more southerly declination.

Both Eratosthenes and Hyginus overlooked the fact that the celestial river is visualized as flowing from north to south, opposite to the direction of the real Nile. Adding to the confusion, later Greek and Latin writers identified the Eridanus with the river Po which flows from west to east across northern Italy.

In mythology, the river Eridanus features in the story of Phaethon, son of the Sun-god Helios, who begged to be allowed to drive his father's chariot across the sky. Reluctantly Helios agreed to the request, but warned Phaethon of the dangers he was facing. 'Follow the track across the heavens where you will see my wheel marks', Helios advised.

As Dawn threw open her doors in the east, Phaethon enthusiastically mounted the Sun-god's golden chariot studded with glittering jewels, little knowing what he was letting himself in for. The four horses immediately sensed the lightness of the chariot with its different driver and they bolted upwards into the sky, off the beaten track, with the chariot bobbing around like a poorly ballasted ship behind them. Even had Phaethon known where the true path lay, he lacked the skill and the strength to control the reins.

The team galloped northwards, so that for the first time the stars of the Plough grew hot and Draco, the dragon, which until then had been sluggish with the cold, sweltered in the heat and snarled furiously. Looking down on Earth from the dizzying heights, the panic-stricken Phaethon grew pale and his knees trembled in fear. Finally, he saw the menacing sight of the Scorpion with its huge claws outstretched and its poisonous tail raised to strike. The swooning Phaethon let the reins slip from his grasp and the horses galloped out of control.

Ovid graphically describes Phaethon's crazy ride in Book II of his *Metamorphoses*. The chariot plunged so low that the Earth caught fire. Enveloped in hot

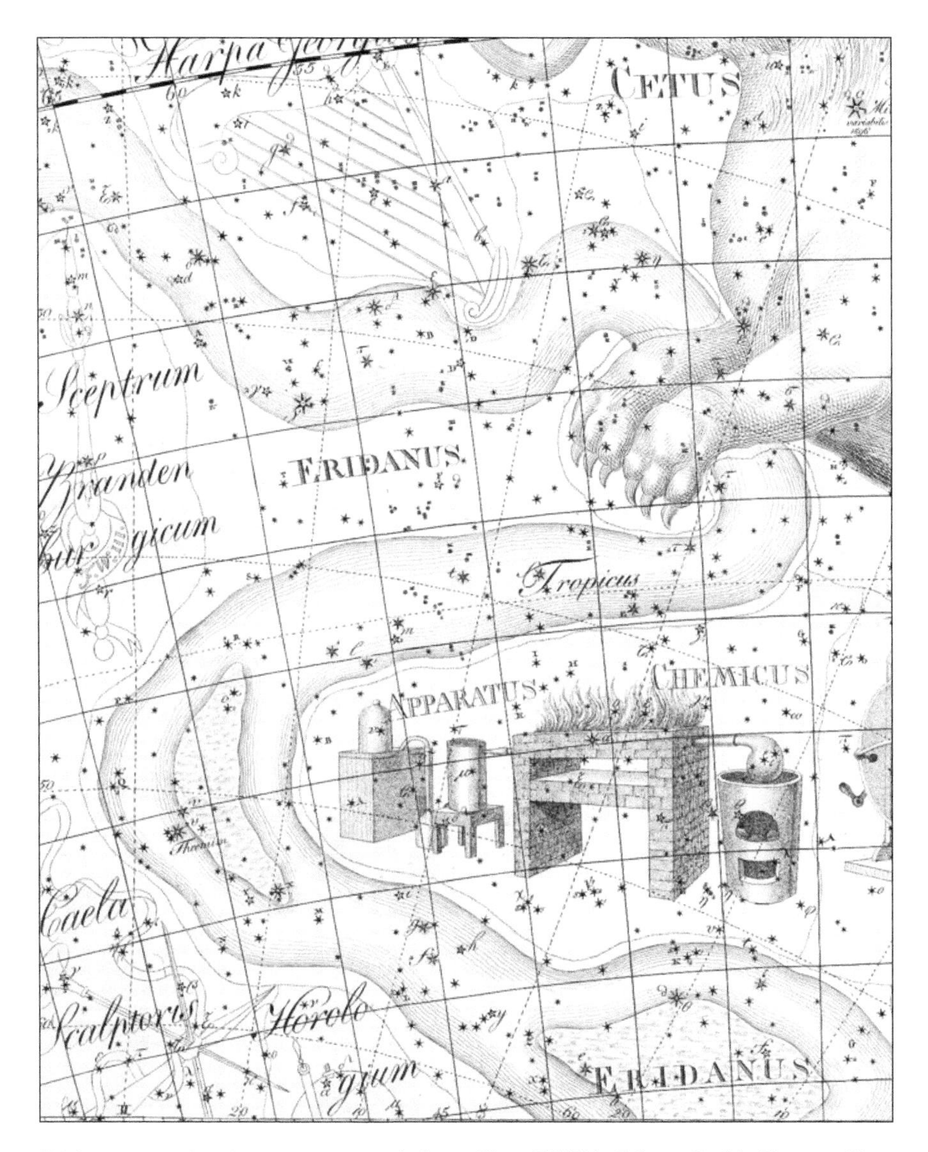

Eridanus meanders from north to south down Chart XVII in Johann Bode's *Uranographia* (1801). At upper right are the flippers of Cetus, and below them lies Apparatus Chemicus, the name given by Bode to the constellation we now know as Fornax. (Deutsches Museum)

smoke, Phaethon was swept along by the horses, not knowing where he was. It was then, the mythologists say, that Libya became a desert, the Ethiopians acquired their dark skins and the seas dried up. To bring the catastrophic events to an end, Zeus struck Phaethon down with a thunderbolt. With his hair streaming fire, the youth plunged like a shooting star into the Eridanus. Some time later, when the Argonauts sailed up the river, they found his body still smouldering, sending up clouds of foul-smelling steam in which birds choked and died. Aratus referred to the 'poor remains' of Eridanus, implying that much of the river's flow was evaporated by the heat of Phaethon's fall.

Eridanus is a long constellation, the sixth-largest in the sky, meandering from the foot of Orion far into the southern hemisphere, ending near Tucana, the toucan. The present-day Eridanus has the greatest north-to-south span of any constellation, nearly 60°. Its brightest star, first-magnitude Alpha Eridani, is called Achernar, from the Arabic *akhir al-nahr* meaning 'the river's end'; at declination −57.2°, it does indeed mark the southern end of Eridanus.

Eridanus extended

In Ptolemy's day, though, the river dried up 17° farther north, at the star Johann Bayer labelled theta (θ). The name Achernar was transferred from this star to its present position when Eridanus was extended south in the late 16th century. Theta Eridani was renamed Acamar, a name that comes from the same Arabic original as Achernar. The present-day Achernar is the only first-magnitude star not listed in Ptolemy's *Almagest*, because it was too far south for him to see.

Eridanus was first shown flowing southwards to the present-day Alpha Eridani on a globe of 1598 compiled by Petrus Plancius. Plancius got his information on the southern stars from observations made by the navigator Pieter Dirkszoon Keyser during the first Dutch voyage to the East Indies in 1595–97. Whether the idea of extending Eridanus was due to Plancius, Keyser, or even some earlier navigators who had previously seen this star is not known.

Bayer showed the southern extension of the river on his *Uranometria* atlas of 1603. It consisted of five stars in all. Bayer labelled them in order of increasing southerly declination with the Greek letters Iota (ι), Kappa (κ), Phi (φ), Chi (χ), and Alpha (α), which they still bear today.

Fornax

The furnace

Genitive: Fornacis
Abbreviation: For
Size ranking: 41st
Origin: The 14 southern constellations of Nicolas Louis de Lacaille

An obscure constellation introduced by the Frenchman Nicolas Louis de Lacaille after his trip to the Cape of Good Hope to observe the southern stars in 1751–52. It lies tucked into a bend in the river Eridanus. Lacaille originally called it le Fourneau on his 1756 planisphere and depicted it as a chemist's furnace used for distillation. He Latinized the name to Fornax Chimiae on the 1763 edition of his planisphere.

It is sometimes said that Lacaille invented the constellation to honour his countryman Antoine Lavoisier, one of the founders of chemistry. This is a misunderstanding, since Lavoisier was only 13 years old when Lacaille's chart of the southern constellations was first published. In fact, the connection with Lavoisier was due to Johann Bode who reinvented the constellation nearly half a century later in his *Uranographia* atlas of 1801. Bode's depiction of Fornax was based on Lavoisier's own diagram of his experiment to decompose water into its constituents of hydrogen and oxygen, as published in *Traité élémentaire de chimie* (1789). As part of his reinvention, Bode retitled the constellation Apparatus Chemicus, although most astronomers continued to use Lacaille's original name.

In 1845 the English astronomer Francis Baily shortened its name to Fornax in his *British Association Catalogue*, acting on a suggestion by John Herschel that all Lacaille's two-word names for constellations should be reduced to one. It has been known as Fornax ever since. Fornax contains no stars brighter than fourth magnitude and none of them is named. For an illustration, see Bode's chart of Eridanus on page 95.

Gemini

The twins

Genitive: Geminorum
Abbreviation: Gem
Size ranking: 30th
Origin: One of the 48 Greek constellations listed by Ptolemy in the *Almagest*
Greek name: Δίδυμοι (Didymoi)

Gemini represents the mythical Greek twins Kastor (Κάστωρ) and Polydeukes (Πολυδεύκης). The Latinized forms of their names are Castor and Pollux (sometimes Polydeuces), by which they are now generally known. The Greeks referred to them jointly as the Dioskouroi (Dioscuri in Latin), literally meaning 'sons of Zeus'. However, mythologists disputed whether both really were sons of Zeus, because of the unusual circumstances of their birth. Their mother was Leda, Queen of Sparta, whom Zeus visited one day in the form of a swan (represented by the constellation Cygnus). That same night she also slept with her husband, King Tyndareus. Both unions were fruitful, for Leda subsequently gave birth to four children. In the most commonly accepted version, Pollux and Helen (later to become famous as Helen of Troy) were children of Zeus, and hence immortal, while Castor and Clytemnestra were fathered by Tyndareus, and hence were mortal.

Castor and Pollux grew up the closest of friends, never quarrelling or acting without consulting each other. They were said to look alike and even to dress alike, as identical twins often do. Castor was a famed horseman and warrior who taught Heracles to fence, while Pollux was a champion boxer.

The inseparable twins joined the expedition of Jason and the Argonauts in search of the golden fleece. The boxing skills of Pollux came in use when the Argonauts landed in a region of Asia Minor ruled by Amycus, a son of

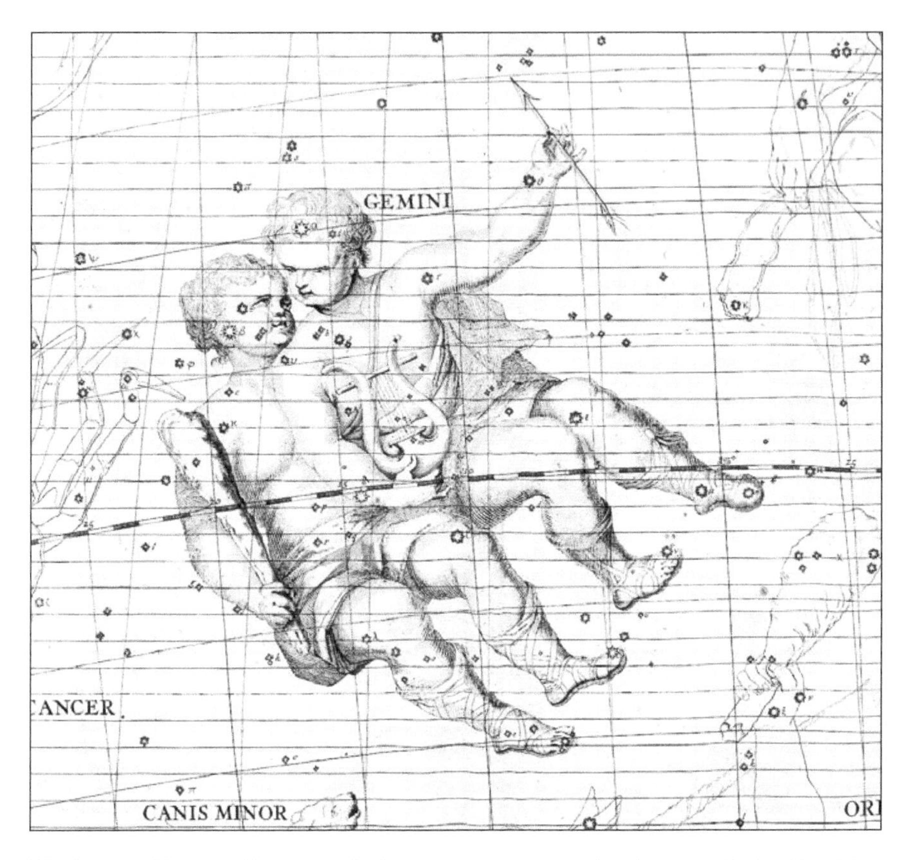

The inseparable twins Castor and Pollux are commemorated in the constellation Gemini, depicted here in the *Atlas Coelestis* of John Flamsteed (1729). Castor carries a lyre and an arrow, Pollux a club. The stars Castor and Pollux mark the heads of the twins. The dashed line cutting through the constellation is the ecliptic. (University of Michigan Library)

Poseidon. Amycus, the world's greatest bully, would not allow visitors to leave until they had fought him in a boxing match, which he invariably won. He stamped down to the shore where the Argo lay and challenged the crew to put up a man against him. Pollux, stirred by the man's arrogance, accepted at once and the two pulled on leather gloves. Pollux easily avoided the rushes of his opponent, like a matador side-stepping a charging bull, and felled Amycus with a blow to the head that splintered his skull.

On the Argonauts' homeward trip with the golden fleece Castor and Pollux were of further value to the crew. Apollonius Rhodius tells us briefly that during the voyage from the mouth of the Rhone to the Stoechades Islands (the present-day Iles d'Hyères off Toulon) the Argonauts owed their safety to Castor and Pollux. Presumably a storm was involved, but he does not elaborate on the circumstances. Ever since this episode, says Apollonius – and he assures us there were other voyages on which they were saviours – the twins have been the patron saints of sailors. Hyginus said that the twins were given the power to save shipwrecked sailors by Poseidon, the sea god, who also presented them with the white horses that they often rode.

Mariners believed that during storms at sea the twins appeared in a ship's rigging in the form of the electrical phenomenon known as St Elmo's fire, as described by Pliny, the first-century Roman writer, in his book *Natural History*:

> On a voyage stars alight on the yards and other parts of the ship. If there are two of them, they denote safety and portend a successful voyage. For this reason they are called Castor and Pollux, and people pray to them as gods for aid at sea.

A single glow was called Helena and was considered a sign of disaster.

Castor and Pollux clashed with another pair of twins, Idas and Lynceus, over two beautiful women. Idas and Lynceus (who were also members of the Argo's crew) were engaged to Phoebe and Hilaira, but Castor and Pollux carried them off. Idas and Lynceus gave pursuit and the two sets of twins fought it out. Castor was run through by a sword thrust from Lynceus, whereupon Pollux killed him. Idas attacked Pollux but was repulsed by a thunderbolt from Zeus.

Another story says that the two pairs of twins made up their quarrel over the women, but came to blows over the division of some cattle they had jointly rustled. Whatever the case, Pollux grieved for his fallen brother and asked Zeus that the two should share immortality. Zeus placed them both in the sky as the constellation Gemini, where they are seen in close embrace, inseparable to the last.

The two brightest stars in the constellation, marking the heads of the twins, are named Castor and Pollux. Astronomers have found that Castor is actually a complex system of six stars linked by gravity, although to the eye they appear as one. Pollux is an orange giant star. Although Castor is labelled Alpha Geminorum, it is actually fainter than Pollux, which is Beta Geminorum. Unlike the twins that they represent, the stars Castor and Pollux are not related since they lie at different distances from us, 52 and 34 light years respectively. Eta Geminorum is called Propus, from the Greek πρόπους (*propous*) meaning 'forward foot', a name that first appears with Eratosthenes; it arises from the star's position in the left (leading) foot of the advance twin, Castor.

Another identification – Apollo and Heracles

Aratus referred to the constellation only as the twins (Δίδυμοι, i.e. Didymoi), without identifying who they were, but a century later Eratosthenes named them as the Dioskouroi, in reference to Castor and Pollux. An alternative view, reported by Hyginus, says that the constellation represents Apollo and Heracles (i.e. Hercules), both sons of Zeus but not twins. Ptolemy called the constellation the Twins (Δίδυμοι) in the *Almagest*, but in a later, more obscure treatise about astrology, called *Tetrabiblos*, he referred to Castor as 'the star of Apollo' and Pollux as 'the star of Heracles', supporting the identifications given by Hyginus.

Several star charts personify the twins as Apollo and Heracles. On the illustration from John Flamsteed's atlas shown here, for example, one twin is depicted holding a lyre and arrow, attributes of Apollo, while the other carries a club, as did Heracles. Bode's *Uranographia* depicts them in the same way.

Grus

The crane

Genitive: Gruis
Abbreviation: Gru
Size ranking: 45th
Origin: The 12 southern constellations of Keyser and de Houtman

One of the 12 constellations introduced at the end of the 16th century by the Dutch navigators Pieter Dirkszoon Keyser and Frederick de Houtman after their pioneer observations of the southern skies. Grus represents a long-necked wading bird, the crane. Possibly they had in mind the sarus crane of India and southeast Asia, which is the largest species of crane, standing nearly 6 ft tall.

The constellation was first shown on a celestial globe by Petrus Plancius and Jodocus Hondius in 1598 under the name Krane Grus, respectively Dutch and Latin words for crane. De Houtman called it Den Reygher, the heron, in his southern star catalogue of 1603, but Johann Bayer adopted the original name Grus for the constellation in his *Uranometria* atlas of 1603.

A second alternative title, Phoenicopterus, the flamingo, first appeared in 1605 in the *Cosmographiae Generalis* of Paul Merula (1558–1607), the librarian of Leiden University, who got his information on the new southern constellations from Plancius. The name Phoenicopterus appeared again on a globe produced around 1625 by the Dutch globe maker Pieter van den Keere (1571–*c.*1646) (Petrus Kaerius in Latin), another Plancius collaborator. However, probably due

Grus cranes its neck to the south of Piscis Austrinus, here called Piscis Notius, on Chart XX of the *Uranographia* of Johann Bode (1801). (Deutsches Museum)

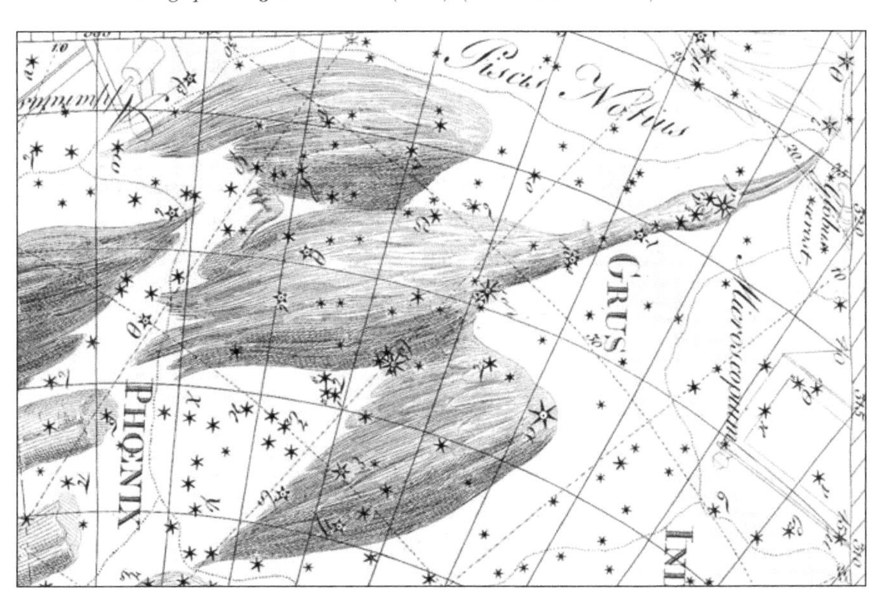

to the influence of Bayer's highly popular atlas, the original identification as a crane won out.

Grus was formed from stars south of Piscis Austrinus, the southern fish. In fact, the star we now know as Gamma Gruis, marking the crane's head, was taken over from the original Greek Piscis Austrinus – Ptolemy described it in the *Almagest* as lying on the tip of the fish's tail. Bayer and others straightened out the tail of Piscis Austrinus so that it did not overlap with the head of Grus.

The constellation's brightest star, second-magnitude Alpha Gruis, is named Alnair, from an abbreviation of the Arabic *al-nayyir min dhanab al-ḥūt* meaning 'the bright one from the fish's tail'. This name arose because Arab astronomers in the 16th century had extended the tail of Piscis Austrinus southwards beyond the Ptolemaic limits of the constellation; it was never part of the original Greek version. There are no legends associated with Grus, but in Greek mythology the crane was sacred to Hermes.

Hercules

Genitive: Herculis
Abbreviation: Her
Size ranking: 5th
Origin: One of the 48 Greek constellations listed by Ptolemy in the *Almagest*
Greek name: 'Εγγόνασι (Engonasi)

The origin of this constellation is so ancient that its true identity was lost even to the Greeks, who knew the figure as either 'Εγγόνασι (Engonasi) or 'Εγγόνασιν (Engonasin), literally meaning 'the kneeling one'. The Greek poet Aratus described him as being worn out with toil, his hands upraised, with one knee bent and a foot on the head of Draco, the dragon. 'No one knows his name, nor what he labours at', said Aratus. Eratosthenes, a century after Aratus, identified the figure as Heracles (the Greek name for Hercules) triumphing over the dragon that guarded the golden apples of the Hesperides. The Greek playwright Aeschylus, quoted by Hyginus, offered a different explanation. He said that Heracles was kneeling, wounded and exhausted, during his battle with the Ligurians.

Heracles is the greatest of Greek and Roman heroes, the equivalent of the Sumerian hero Gilgamesh. So it is surprising that the Greeks allotted him a constellation only as an afterthought. One reason may be that he was already sometimes personified as one of the heavenly twins represented by the constellation Gemini, the other twin being Apollo.

The full saga of Heracles is long and complex, as befits a legend that has grown in the telling. Heracles was the illicit son of the god Zeus and Alcmene, most beautiful and wise of mortal women, whom Zeus visited in the form of her husband, Amphitryon. The infant was christened Alcides, Alcaeus, or even Palaemon, according to different accounts; the name Heracles came later. Zeus's real wife, Hera, was furious at her husband's infidelity. Worse still, Zeus laid the

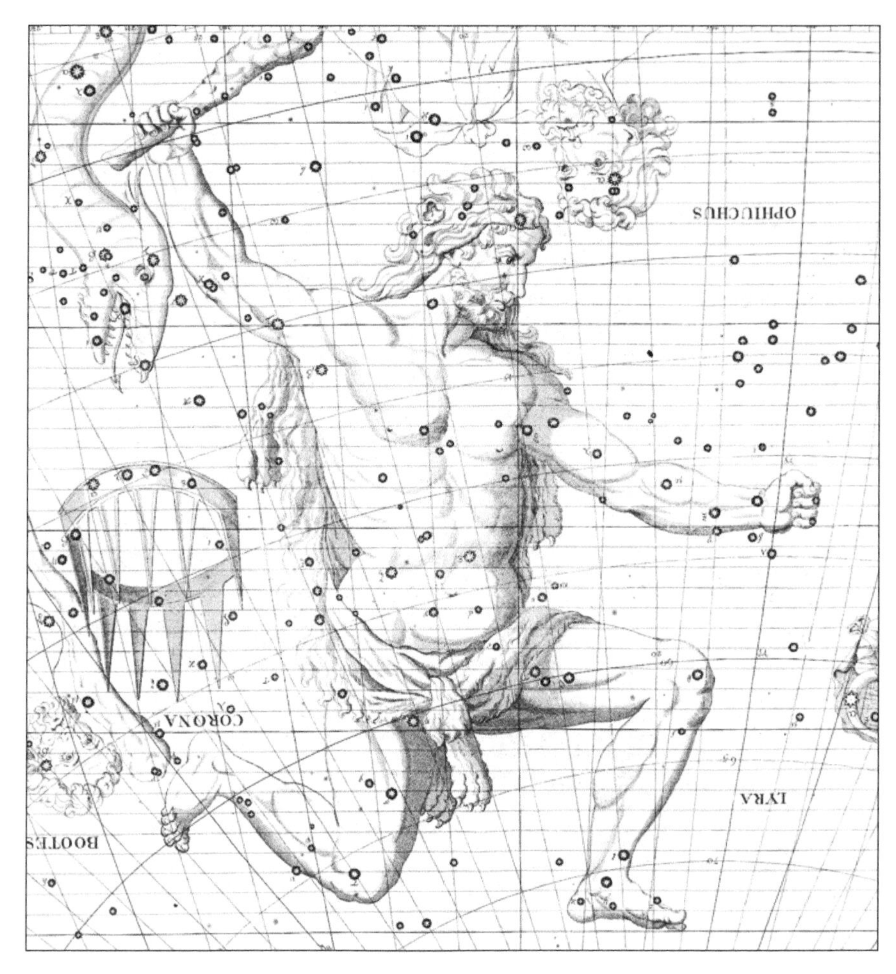

Hercules, the kneeling man, from the *Atlas Coelestis* of John Flamsteed (1729). In the sky he is depicted with his feet towards the north celestial pole, his left foot on the head of Draco, the dragon. Hercules wears a lion's skin and in his right hand brandishes a club, his favourite weapon. Here his left hand is empty, but other illustrations show it grasping either the three-headed Cerberus (p. 190) or an apple branch. (University of Michigan Library)

infant Heracles at Hera's breast while she slept, so that he suckled her milk. And having drunk the milk of a goddess, Heracles became immortal.

As Heracles grew up he surpassed all other men in size, strength and skills with weapons, but he was forever dogged by the jealousy of Hera. She could not kill him, since he was immortal, so instead she vowed to make his life as unpleasant as possible. Under Hera's evil spell he killed his children in a fit of madness. When sanity returned, he went remorsefully to the Oracle at Delphi to ask how he might atone for his dreadful deed. The Oracle ordered him to serve Eurystheus, king of Mycenae, for 12 years. It was then that the Oracle gave him the name Heracles, meaning 'glory of Hera'.

Labours of Heracles

Eurystheus set him a series of ten tasks that are called the Labours of Heracles. The first was to kill a lion that was terrorizing the land around the city of Nemea. This lion had a hide that was impervious to any weapon – so Heracles strangled it to death. He used its own claws to cut off the skin. Thereafter he wore the pelt of the lion as a cloak, with its gaping mouth as a helmet, which made him look even more formidable. The Nemean lion is identified with the constellation Leo.

The second labour was to destroy the multi-headed monster called the Hydra which lurked in the swamp near the town of Lerna, devouring incautious passers-by. Heracles grappled with the monster, but as soon as he cut off one of its heads, two grew to replace it. To make matters worse, a large crab came scuttling out of the swamp and nipped at the feet of Heracles. Angrily he stamped on the crab and called for help to Iolaus, his charioteer, who burned the stumps as each head was lopped to prevent more heads growing. Heracles gutted the Hydra and dipped his arrows in its poisonous blood – an action that would eventually be his undoing. Both the crab (Cancer) and the Hydra are commemorated as constellations.

For his next two labours, Heracles was ordered to catch elusive animals: a deer with golden horns, and a ferocious boar. Perhaps the most famous labour is his fifth, the cleaning of the dung-filled stables of King Augeias of Elis. Heracles struck a bargain with the king that he would clean out the stables in a single day in return for one-tenth of the king's cattle. Heracles accomplished the task by diverting two rivers. But Augeias, claiming he had been tricked, renounced the bargain, and banished Heracles from Elis.

The next task took him to the town of Stymphalus where he dispersed a flock of marauding birds with arrow-like feathers. The survivors flew to the Black Sea, where they subsequently attacked Jason and the Argonauts. Next, Heracles sailed to Crete to capture a fire-breathing bull that was ravaging the land. Some equate this bull with the constellation Taurus. For his eighth and ninth labours, Heracles brought to Eurystheus the flesh-eating horses of King Diomedes of Thrace and the belt of Hippolyte, queen of the Amazons.

Finally, Heracles was sent to steal the cattle of Geryon, a triple-bodied monster that ruled the island of Erytheia, far to the west. While sailing there, Heracles set up the columns at the straits of Gibraltar called the Pillars of Heracles. He killed Geryon with a single arrow that pierced all three bodies from the side, then drove the cattle back to Greece. Passing through Liguria, in southern France, he was set upon by local forces who so outnumbered him that he ran out of arrows. Sinking to his knees, he prayed to his father, Zeus, who rained down rocks on the plain. Heracles hurled the rocks at his attackers and routed them. According to Aeschylus, this is the incident that is recorded by the constellation Engonasin, the kneeler.

Two more tasks – mission creep

When Heracles returned from the last of these exploits, the cowardly and deceitful Eurystheus refused to release him from his service because Heracles had received help in slaying the Hydra and had attempted to profit from the stable-cleaning. Hence Eurystheus set two additional tasks, even more difficult than those before. The first was to steal the golden apples from the garden of Hera on the slopes of Mount Atlas. The tree with the golden fruit had been a wedding present from Mother Earth (Gaia) when Hera married Zeus. Hera set the Hesperides, daughters of Atlas, to guard the tree, but they stole some of the precious produce. So now the dragon Ladon lay coiled around the tree to prevent any further pilfering.

After a heroic journey, during which he released Prometheus from his bonds, Heracles came to the garden where the golden apples grew. Nearby stood Atlas, supporting the heavens on his shoulders. Heracles dispatched Ladon with a well-aimed arrow, and Hera set the dragon in the sky as the constellation Draco. Heracles had been advised (by Prometheus, says Apollodorus) not to pick the apples himself, so he invited Atlas to fetch them for him while he temporarily supported the skies. Heracles hastily returned the burden of the skies to the shoulders of Atlas before making off with the golden treasure.

The twelfth labour, the most daunting of all, took him down to the gates of the Underworld to fetch Cerberus, the three-headed watchdog. Cerberus had the tail of a dragon and his back was covered with snakes. A more loathsome creature would be difficult to imagine but Heracles, protected from the tail and the snakes by the skin of the Nemean lion, wrestled Cerberus with his bare hands and dragged the slavering dog to Eurystheus. The startled king had never expected to see Heracles alive again. Now, with all the labours completed, Eurystheus had no option but to make Heracles a free man again.

Death of Heracles

The death of Heracles is a true piece of Greek tragedy. After his labours, Heracles married Deianeira, the young and beautiful daughter of King Oeneus. While travelling together, Heracles and Deianeira came to the swollen river Evenus where the centaur Nessus ferried passengers across. Heracles swam across himself, leaving Deianeira to be carried by Nessus. The centaur, aroused by her beauty, tried to ravish her, and Heracles shot him with one of his arrows tipped with the Hydra's poison.

The dying centaur offered Deianeira some of his blood, deceitfully claiming that it would act as a love charm. Innocently, Deianeira accepted the poisoned blood and kept it safely until, much later, she began to suspect that Heracles had his eye on another woman. In the hope of rekindling his affection, Deianeira gave Heracles a shirt on which she had smeared the blood of the dying Nessus. Heracles put it on – and as the blood warmed up, the Hydra's poison began to burn his flesh to the bone.

In agony, Heracles raged over the countryside, tearing up trees. Realizing there was no release from the pain, he built himself a funeral pyre on Mount Oeta, spread out his lion's skin and lay down on it, peaceful at last. The flames burned up the mortal part of him, while the immortal part ascended to join the gods on Mount Olympus. His father, Zeus, turned him into a constellation, which we know by the Latin name Hercules.

Heracles is depicted in the sky holding a club, his favourite weapon. Some people think that his 12 labours are represented by the 12 signs of the zodiac, but it is difficult to see the connection in some cases.

Stars of Hercules

Hercules is the fifth-largest constellation but is not particularly prominent. Alpha Herculis, a red giant star that varies from third to fourth magnitude, is called Rasalgethi, from the Arabic meaning 'the kneeler's head', which it marks. Beta and Delta Herculis, named Kornephoros and Sarin, are his right and left shoulders respectively and his left arm extends towards Lyra. The four stars Epsilon, Zeta, Eta, and Pi Herculis form a distinctive quadrilateral known as the Keystone that outlines his pelvis. In some depictions, such as in Johann Bayer's *Uranometria*, Hercules was imagined holding a branch from the apple tree of the Hesperides in his left hand (see page 32). Johannes Hevelius, in his own atlas, replaced the apple branch with the three-headed monster Cerberus (see page 190).

His left leg, with Theta Herculis as the knee and Iota Herculis as the lower shin, presses on the head of the vanquished Draco, the dragon. Hercules rests on his right knee, represented by Tau Herculis. In Ptolemy's day the star we now know as Nu Boötis doubled up as the sole of his right foot, in an example of stars being shared by neighbouring constellations.

Astronomically speaking, the most celebrated object in the constellation is a globular cluster of stars, M13, the best example of such a cluster in northern skies. It was discovered by chance in 1714 by Edmond Halley when he was Professor of Geometry at Oxford. He described it as 'a little patch, but it shews it self to the naked eye when the sky is serene and the Moon absent'. Whether he actually discovered it with the naked eye or telescopically he did not say.

Horologium

The pendulum clock

Genitive: Horologii
Abbreviation: Hor
Size ranking: 58th
Origin: The 14 southern constellations of Nicolas Louis de Lacaille

One of the small southern constellations introduced by the Frenchman Nicolas Louis de Lacaille after he mapped the southern stars in 1751–52. Lacaille wrote that the constellation represented a pendulum clock beating seconds, as used for timing his observations. Lacaille introduced it on his first chart in 1756 under the French name l'Horloge, but this was Latinized to Horologium on the second edition in 1763.

The clock was depicted with a fully marked dial and even a seconds-hand, a remarkable feat of imagination for an area of sky that contains only a sparse

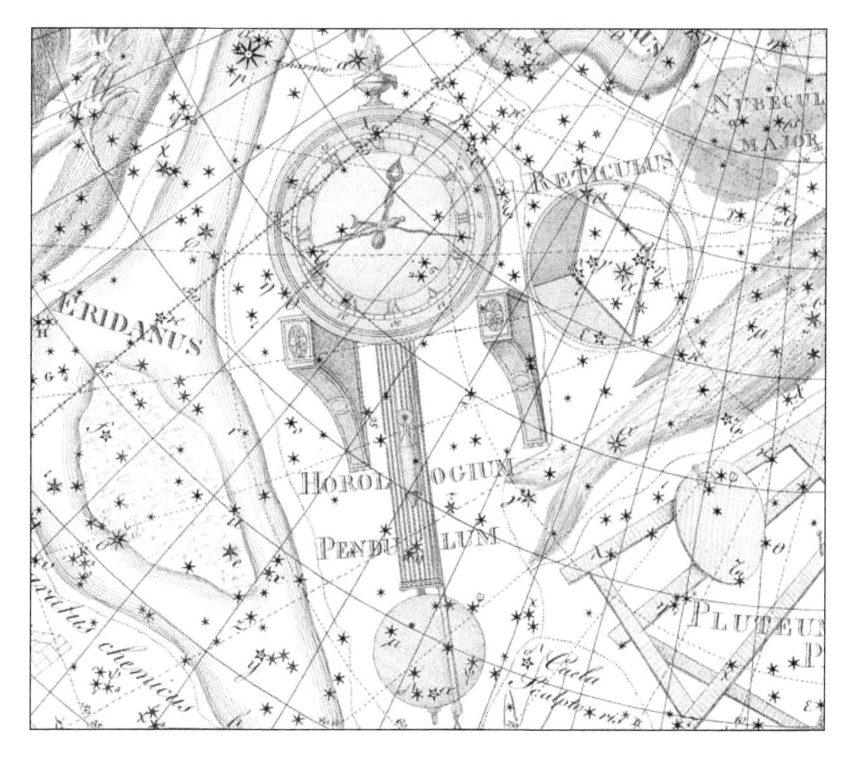

Horologium, under the name Horologium Pendulum, on Chart XX of Johann Bode's *Uranographia* star atlas (1801). Bode showed it with a nine-rod gridiron pendulum, as invented *c.*1726 by the English clockmaker John Harrison. (Deutsches Museum)

scattering of stars no brighter than fourth magnitude. In some representations its brightest star, Alpha Horologii, marks the bob on the end of the clock's pendulum, as in Bode's illustration above, while others such as Lacaille himself placed Alpha on one of the driving weights.

Hydra

The water snake

Genitive: Hydrae
Abbreviation: Hya
Size ranking: 1st
Origin: One of the 48 Greek constellations listed by Ptolemy in the *Almagest*
Greek name: Ὕδρος (Hydros)

Hydra is the largest of the 88 constellations, winding over a quarter of the way around the sky. Its head is south of the constellation Cancer, the crab, while the tip of its tail lies between Libra, the scales, and Centaurus, the centaur.

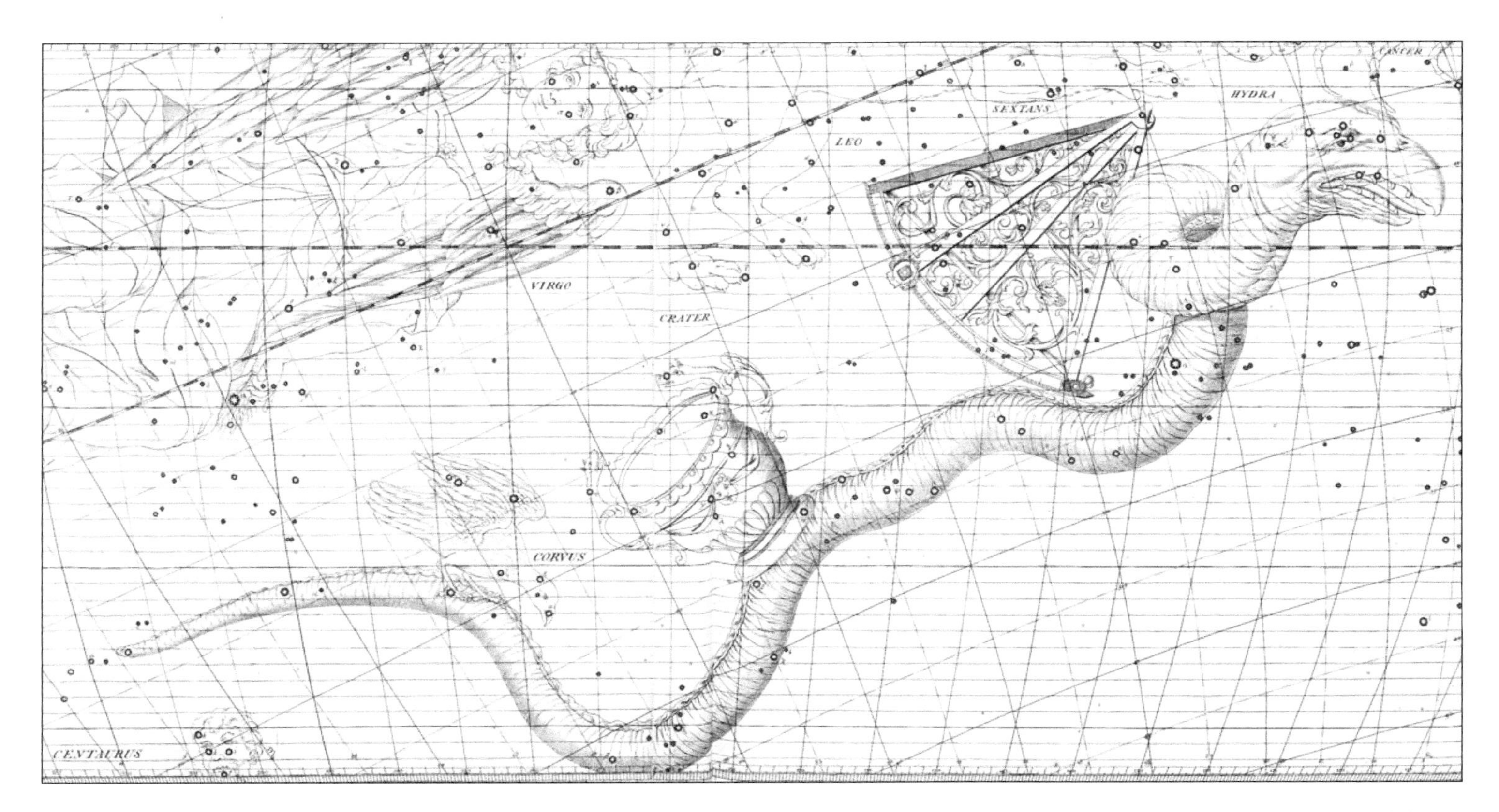

Hydra winds across the pages of John Flamsteed's *Atlas Coelestis* (1729). On its back are Corvus and Crater, associated with it in legend. Because of the constellation's extreme length, fold-outs were required to fit it on one chart. (University of Michigan Library)

The total length from its westernmost boundary to the easternmost one is 102.5°. Yet for all its size there is nothing prominent about Hydra. Its only star of note is second-magnitude Alphard, a name that comes from the Arabic *al-fard* appropriately meaning 'the solitary one'.

Eratosthenes and Ptolemy both called the constellation Ὕδρος (Hydros), which is the male form of the name; but this water snake is actually female, so they were in error. Aratus correctly used the feminine form, Ὕδρη (Hydra). There is now also a male water snake in the sky, the small southern constellation Hydrus, introduced by Dutch navigators at the end of the 16th century.

Heracles fights the Hydra

The water-snake features in two legends. First, and most familiar, the Hydra was the creature that Heracles fought and killed as the second of his famous labours. The Hydra was a multi-headed creature, the offspring of the monster Typhon and the half-woman, half-serpent called Echidna. Hydra was thus the sister of the dragon that guarded the golden apples, commemorated by the constellation Draco. Hydra reputedly had nine heads, the middle one of which was immortal. In the sky, though, it is shown with one head only – perhaps this is the immortal one.

Hydra lived in a swamp near the town of Lerna, from where it sallied forth over the surrounding plain, eating cattle and ravaging the countryside. Its breath and even the smell of its tracks were said to be so poisonous that anyone who breathed them died in agony.

Heracles rode up to the Hydra's lair in his chariot and fired flaming arrows into the swamp to force the creature into the open, where he grappled with it. The Hydra wrapped itself around one of his legs; Heracles smashed at its heads with his club but no sooner had one head been destroyed than two grew in its place. To add to Heracles's worries, a huge crab scuttled out of the swamp and attacked his other foot, but Heracles stamped on the crab and crushed it. The crab is commemorated in the constellation Cancer.

Heracles called for help to his charioteer Iolaus who burned the stump of each head as soon as it was struck off to prevent others growing in its place. Finally Heracles cut off the immortal head of the Hydra and buried it under a heavy rock by the roadside. He slit open the body of the Hydra and dipped his arrows in its poisonous gall.

The crow and the cup

A second legend associates the water-snake with the constellations of the Crow (Corvus) and the Cup (Crater) that lie on its back. In this story, the crow was sent by Apollo to fetch water in the bowl, but loitered to eat figs from a tree. When the crow eventually returned to Apollo it blamed the water-snake for blocking the spring. But Apollo knew that the crow was lying, and punished the bird by placing him in the sky, where the water-snake eternally prevents him from drinking out of the bowl. For more about the constellations Crater and Corvus see pages 82–83.

Hydrus

The lesser water snake

Genitive: Hydri
Abbreviation: Hyi
Size ranking: 61st
Origin: The 12 southern constellations of Keyser and de Houtman

A small southern counterpart of the great water-snake, Hydra, with which it is not to be confused. This is one of several examples of the repetition of constellation figures in the sky, as in the Great and Little Bear, the Great and Little Dog, the two lions, the horses Pegasus and Equuleus, the Northern and Southern Crown, and the Northern and Southern Triangle.

Hydrus was one of the 12 southern constellations introduced at the end of the 16th century by the Dutch navigators Pieter Dirkszoon Keyser and Frederick de Houtman and first appeared on Petrus Plancius's globe of 1598. It represents the sea snakes that the Dutch explorers would have seen on their voyages. Hydrus is a male water snake, whereas the much larger Greek constellation Hydra is a female. To emphasize the difference in gender, Nicolas Louis de Lacaille termed it l'Hydre Mâle on his planisphere of the southern skies published in 1756.

Hydrus has endured more redesigns than any other constellation. Originally it was visualized as wriggling beneath the feet of Tucana and Pavo, then curling past the south celestial pole before ending next to Apus, as seen on Johann Bayer's *Uranometria* atlas of 1603. This depiction was based on the now-lost star

Hydrus as shown by Johann Bode on Chart XX of his *Uranographia* (1801). The object labelled Nubecula Minor, at centre, is the Small Magellanic Cloud. Part of the Large Magellanic Cloud, Nubecula Major, is visible at bottom right. (Deutsches Museum)

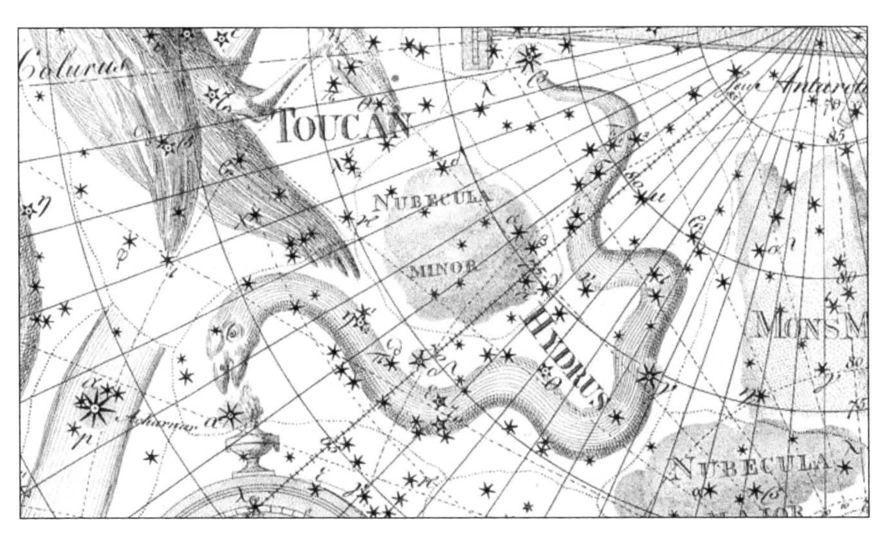

list of Keyser. In de Houtman's slightly later catalogue published in 1603, the tip of the tail did not extend as far south, ending at the star we now know as Nu Octantis.

More severe changes were to come during Lacaille's reorganization of the southern skies a century and a half later. He rerouted Hydrus to pass between the two Magellanic Clouds, transferring some of its stars to Tucana in the process (including the 'star' now known as the globular cluster 47 Tucanae). In addition, Lacaille docked the snake's tail to make way for Octans, one of his own inventions. He also commandeered a couple of stars from Hydrus for Horologium and Reticulum, another two of his new figures. Lacaille's truncated version of Hydrus terminated at Beta Hydri, as shown by Bode (see previous page). It is this more compact snake that we see in the sky today. The brightest stars of Hydrus are of third magnitude, but none are named.

Indus

The Indian

Genitive: Indi
Abbreviation: Ind
Size ranking: 49th
Origin: The 12 southern constellations of Keyser and de Houtman

Indus is one of the 12 figures formed by the Dutch navigators Pieter Dirkszoon Keyser and Frederick de Houtman from stars they charted in the southern hemisphere on their voyages to the East Indies at the end of the 16th century.

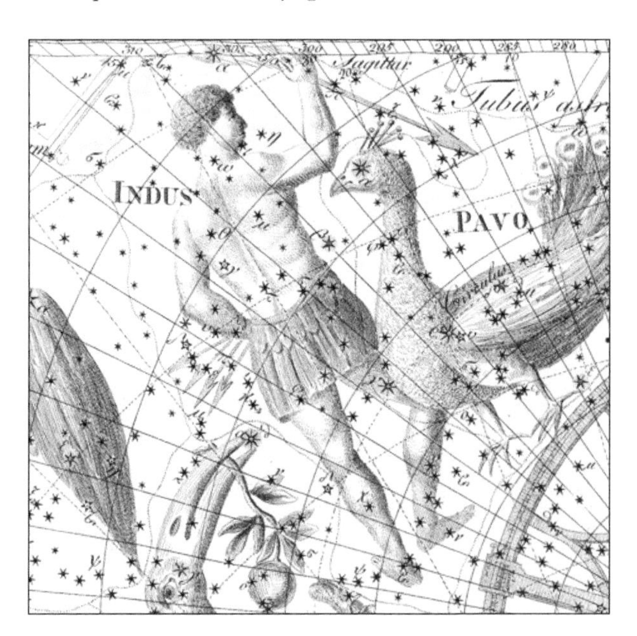

Indus, an Indian brandishing a spear in one hand and holding a clutch of spears or arrows with the other, as shown on Chart XX of the *Uranographia* of Johann Bode (1801). Here it is overlapped by Pavo, the peacock, another new southern constellation. (Deutsches Museum)

It first made its appearance in 1598 on a globe by the Dutch cartographer Petrus Plancius and then first appeared in print in 1603 on the *Uranometria* atlas of Johann Bayer.

The Indian is portrayed brandishing a spear as though hunting. Whether he is supposed to be a native of Madagascar, where the Dutch fleet stayed for several months on their way east and made many of their astronomical observations, is not known. Alternatively he could be a native of the East Indies or southern Africa. Or perhaps the figure is symbolic, representing all the indigenous peoples the explorers encountered on their various travels, from South America to the Indies. The constellation's brightest stars are of third magnitude, but none of them is named.

Lacerta
The lizard

Genitive: Lacertae
Abbreviation: Lac
Size ranking: 68th
Origin: The seven constellations of Johannes Hevelius

This inconspicuous constellation, sandwiched between Cygnus and Andromeda like a lizard between rocks, was introduced by the Polish astronomer Johannes Hevelius in his star catalogue of 1687. Hevelius gave it the alternative title of

Lacerta the celestial lizard slithers across the sky as seen on Chart IV in the *Uranographia* of Johann Bode (1801). (Deutsches Museum)

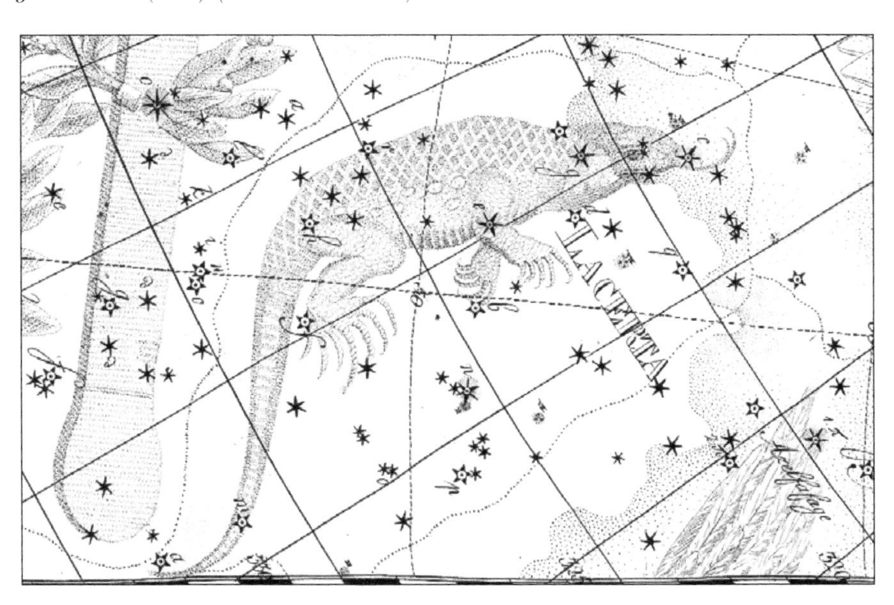

Stellio, a type of lizard also known as a starred agama, but this secondary name soon fell into disuse.

Lacerta was extended northwards by the English astronomer John Flamsteed in his *Catalogus Britannicus* of 1725; he incorporated into it a triangle of stars that Hevelius had depicted as part of the turban of Cepheus. Later mapmakers such as Bode showed these three stars as the head of the enlarged lizard.

Alpha and Beta Lacertae are the constellation's two brightest stars, but are of only fourth magnitude. They are the only two stars in Lacerta with Greek letters; these letters were allocated not by Hevelius but by Francis Baily in his *British Association Catalogue* of 1845. None of the stars of Lacerta have names, nor are there any legends associated with the constellation.

Leo

The lion

Genitive: Leonis
Abbreviation: Leo
Size ranking: 12th
Origin: One of the 48 Greek constellations listed by Ptolemy in the *Almagest*
Greek name: Λέων (Leon)

Eratosthenes and Hyginus both affirm that the lion was placed in the sky because it is the king of beasts. Mythologically speaking, this is reputed to be the lion of Nemea, slain by Heracles as the first of his 12 labours. Nemea is a town some way south-west of Corinth. There the lion lived in a cave with two mouths, emerging to carry off the local inhabitants, who were becoming scarce. The lion was an invulnerable beast of uncertain parentage; it was variously said to have been sired by the dog Orthrus, the monster Typhon, or even to be the offspring of Selene, the Moon goddess. Its skin was proof against all weapons, as Heracles found when he shot an arrow at the lion and saw that it simply bounced off.

Undeterred, Heracles heaved up his mighty club and made after the animal, which retreated into its cave. Heracles blocked up one of the entrances and went in through the other. He grappled with the lion, locking his huge arm around its throat and choking the beast to death. Heracles carried the lion's corpse away in triumph on his shoulders. Later he used the creature's own razor-sharp claws to cut off its pelt, which he wore as a cloak. The lion's gaping mouth bobbing above his own head made Heracles look more fearsome than ever.

The Greeks knew the constellation as Λέων, simply 'lion', and it is easy to make out the shape of a crouching lion among its stars. The head and chest are outlined by six stars arranged in the shape of a sickle. At the foot of the Sickle, marking the lion's heart (according to Ptolemy's description), is the constellation's brightest star, Alpha Leonis, which we call Regulus, Latin for 'little king'. Ptolemy in the *Almagest* called it Βασιλίσκος (Basiliskos, or Basiliscus in Latin

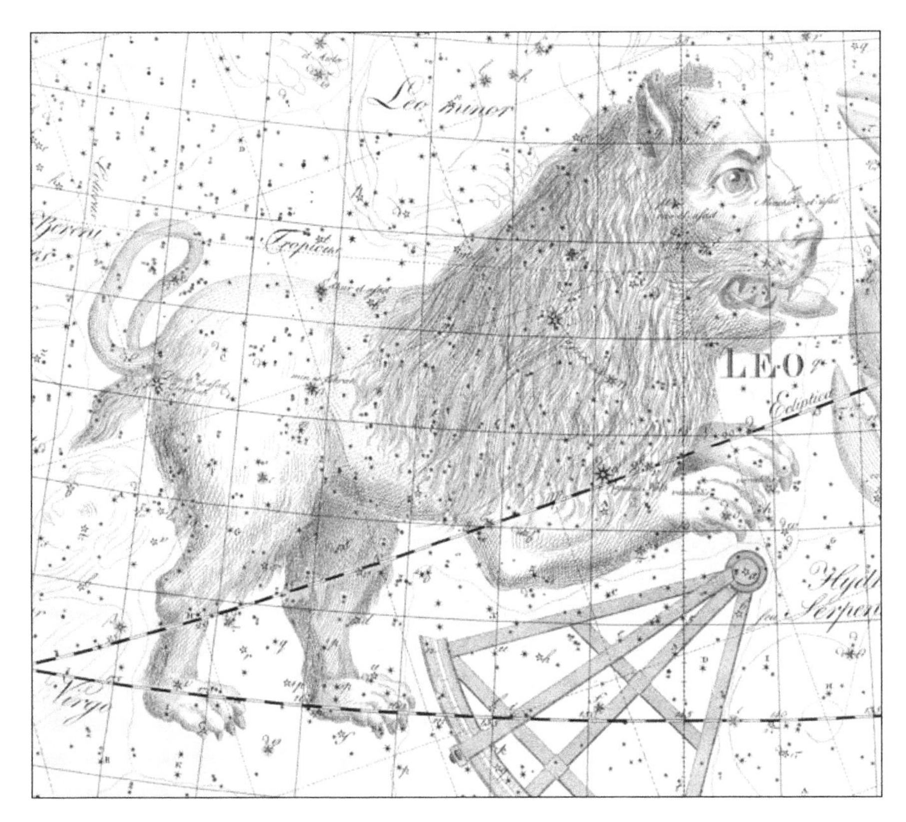

Leo shown ready to pounce on Chart XIII of the *Uranometria* of Johann Bode (1801). In the lion's chest can be found the bright star Regulus, labelled Alpha. Leo lies on the Sun's path around the sky, the ecliptic, here marked by a dashed line. (Deutsches Museum)

transliteration) which means the same. The name Basiliskos first appeared in print with the Greek writer Geminus roughly two centuries before Ptolemy; the earlier writers Aratus and Eratosthenes gave the star no name at all. The Babylonians knew the star as LUGAL, 'king'.

On the tip of the tail is the star Beta Leonis, called Denebola from the Arabic for 'the lion's tail'. Gamma Leonis is called Algieba, from the Arabic meaning 'the forehead'; this seems puzzling, since according to Ptolemy it lies in the lion's neck, but the Arabs saw here a very much larger lion than the one visualized by the Greeks. Gamma Leonis is a celebrated double star, consisting of a pair of yellow giant stars divisible in small telescopes. Delta Leonis is called Zosma from a Greek word meaning 'girdle' or 'loin cloth', mistakenly applied to this star in Renaissance times; in fact it lies in the lion's rump.

Ptolemy listed eight 'unformed' stars lying outside the body of Leo. Three of them formed a triangle to the north of the lion's tail, marking the corners of what Ptolemy referred to as a 'nebulous mass'. This was the large open cluster we know as Melotte 111, now part of Coma Berenices which was made into a separate constellation in the 16th century (page 78).

Leo Minor
The little lion

Genitive: Leonis Minoris
Abbreviation: LMi
Size ranking: 64th
Origin: The seven constellations of Johannes Hevelius

A lion cub accompanying Leo, introduced by the Polish astronomer Johannes Hevelius in his star catalogue and atlas of 1687. He formed it from 18 faint stars between Ursa Major and Leo where the short-lived constellation Jordanus once flowed (page 195). The brightest stars of Leo Minor are of only fourth magnitude and there are no legends associated with it.

Curiously, Leo Minor has no star labelled Alpha, although there is a Beta Leonis Minoris. This seems to have resulted from an oversight on the part of the 19th-century English astronomer Francis Baily. Hevelius did not label the stars in any of his newly formed constellations, so 150 years later Baily did it for him. In his *British Association Catalogue* of 1845, Baily assigned the letter Beta to the second-brightest star in Leo Minor, but left the brightest star (46 Leonis Minoris, on the rump of the lion) unlettered by mistake.

There is further confusion regarding the naming of the brightest star in this constellation. In his book *Star Names, Their Lore and Meaning*, R. H. Allen says that Hevelius described the brightest star in Leo Minor as Praecipua, meaning

Leo Minor, the lion cub, lies immediately above the head of Leo itself, as shown on the *Atlas Coelestis* of John Flamsteed (1729). (University of Michigan Library)

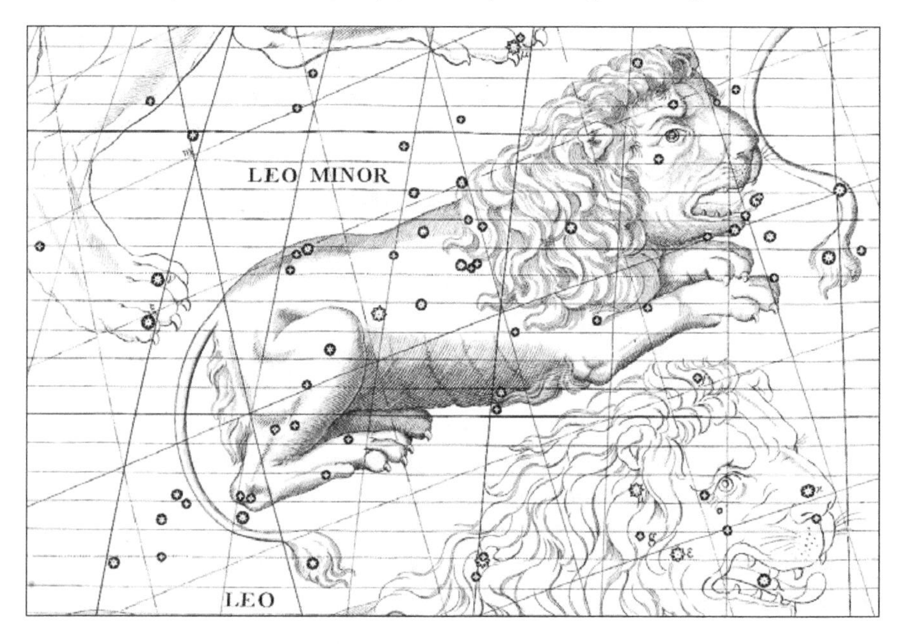

'chief', and that it was later used as a star name by the Italian astronomer Giuseppe Piazzi in his *Palermo Catalogue* of 1814. However, I am unable to find any such mention of a Praecipua in Hevelius's catalogue, although the name does appear in Piazzi's catalogue. Allen also states that the star named Praecipua by Piazzi was 46 LMi, but again he is wrong; Piazzi's Praecipua was actually 37 LMi, which he wrongly assessed as being brighter than 46 LMi. Reasonably enough, modern sources have followed Allen by using the name Praecipua for the brightest star, 46 LMi, rather than the fainter 37 LMi.

Lepus
The hare

Genitive: Leporis
Abbreviation: Lep
Size ranking: 51st
Origin: One of the 48 Greek constellations listed by Ptolemy in the *Almagest*
Greek name: Λαγωός (Lagoös)

The Greeks knew this constellation as Λαγωός (Lagoös), their word for hare; Lepus is the more recent Latin name. Eratosthenes tells us that Hermes placed the hare in the sky because of its swiftness. Both Eratosthenes and Hyginus referred to the remarkable fertility of hares, as attested to by Aristotle in his

Lepus cowers under the feet of Orion, the hunter, as depicted on the *Atlas Coelestis* of John Flamsteed (1729). (University of Michigan Library)

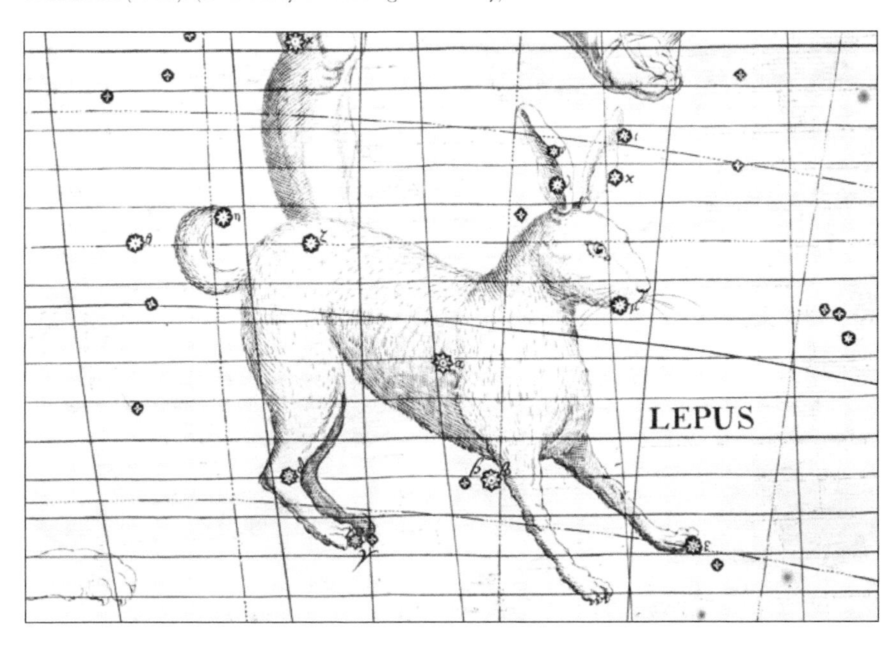

Historia Animalium (*History of Animals*): 'Hares breed and bear at all seasons, superfoetate (i.e. conceive again) during pregnancy and bear young every month.'

The celestial hare makes an interesting tableau with Orion and his dogs. Aratus wrote that the Dog (Canis Major) pursues the hare in an unending race: 'Close behind he rises and as he sets he eyes the setting hare.' But judging by its position in the sky, the hare seems more to be crouched in hiding beneath the hunter's feet.

Hyginus tells us the following moral tale about the hare. At one time there were no hares on the island of Leros, until one man brought in a pregnant female. Soon, everyone began to raise hares and before long the island was swarming with them. They overran the fields and destroyed the crops, reducing the population to starvation. By a concerted effort, the inhabitants drove the hares out of their island. They put the image of the hare among the stars as a reminder that one can easily end up with too much of a good thing.

The constellation's brightest star, third-magnitude Alpha Leporis, is called Arneb, from the Arabic *al-arnab* meaning 'the hare'. It lies in the middle of the animal's body. The stars Kappa, Iota, Lambda, and Nu Leporis delineate the hare's prominent ears.

Libra

The scales

Genitive: Librae
Abbreviation: Lib
Size ranking: 29th
Origin: One of the 48 Greek constellations listed by Ptolemy in the *Almagest*
Greek name: Χηλαί (Chelae)

In ancient Greek times, the area of sky we know as Libra was occupied by the claws of the scorpion, Scorpius. The Greeks called this area Χηλαί (Chelae), literally meaning 'claws', an identification that lives on in the names of the individual stars of Libra (see below). As things have worked out, Libra is now a slightly larger constellation than Scorpius, but is much less conspicuous.

The identification of this area with a balance became established in the first century BC among the Romans, although exactly when it was introduced and by whom has been lost in the mists of history. Ptolemy in the *Almagest*, written around AD 150, continued to refer to this constellation as the Claws, preferring to follow Greek tradition even though it was by then being superseded – for example, Libra was shown as a pair of scales on the celestial globe held by the Farnese Atlas, a Roman statue produced about the same time that the *Almagest* was being written (see page 27 and Fig. 3).

To the Romans, Libra was a favoured constellation. The Moon was said to have been in Libra when Rome was founded. 'Italy belongs to the Balance, her rightful sign. Beneath it Rome and her sovereignty of the world were founded',

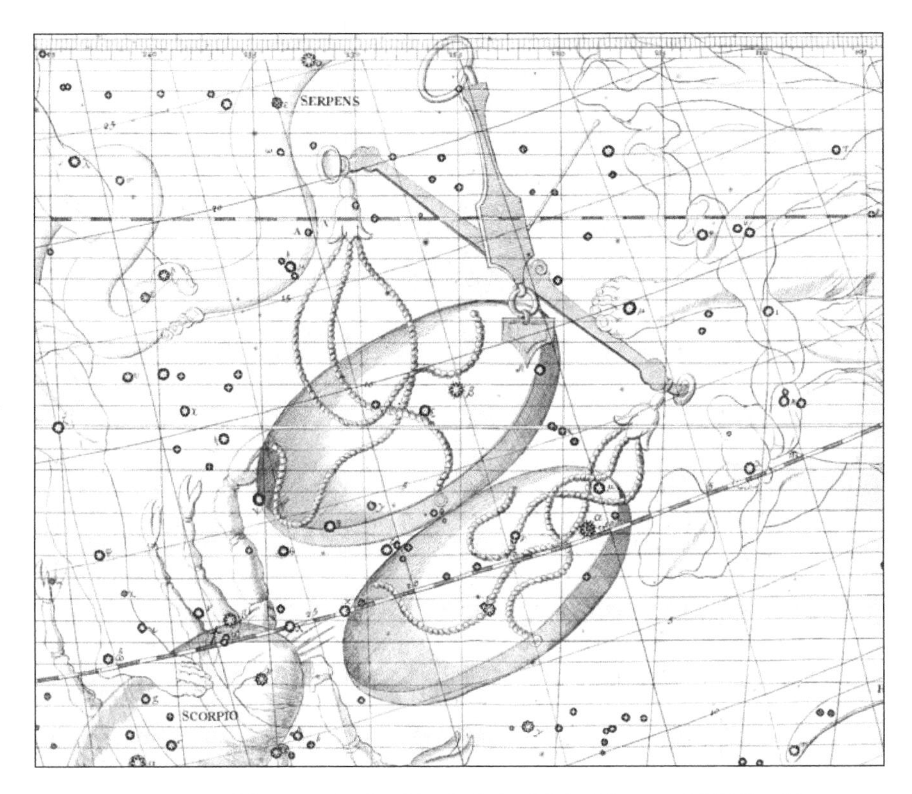

The balance pans of Libra seen in the *Atlas Coelestis* of John Flamsteed (1729), with Scorpius at lower left and the feet of Virgo to the right. (University of Michigan Library)

said the Roman writer Manilius. He described Libra as 'the sign in which the seasons are balanced, and the hours of night and day match each other'. This is a hint that the Romans visualized Libra as a balance because the Sun lay within it at the autumnal equinox, when day and night are equal. However, the idea of a balance in this area did not originate with the Romans. The Babylonians knew this area as ZIB.BA.AN.NA, the balance of heaven, around 1000 BC, when the autumnal equinox lay among its stars. Hence it seems that the Romans revived a constellation that existed even before Greek times.

Libra is the only constellation of the zodiac to represent an inanimate object; the other 11 zodiacal constellations represent animals or mythological characters. Once the identification of Libra with a pair of scales became established it was natural to divorce it entirely from Scorpius and to associate it instead with the other flanking zodiacal figure, Virgo, who was identified with Dike or Astraeia, goddess of justice. Libra thus became the scales of justice held aloft by the goddess – although, in the sky, they were positioned at her feet.

Libra's brightest star, second-magnitude Alpha Librae, is called Zubenelgenubi from the Arabic meaning 'the southern claw', a reminder of the Greek identification of this constellation with the claws of the scorpion. Beta Librae is Zubeneschamali, 'the northern claw'.

Lupus
The wolf

Genitive: Lupi
Abbreviation: Lup
Size ranking: 46th
Origin: One of the 48 Greek constellations listed by Ptolemy in the *Almagest*
Greek name: Θηρίον (Therion)

The ancient Greeks called this constellation Θηρίον (Therion), representing an unspecified wild animal, while the Romans called it Bestia, the beast. It was visualized as impaled on a long pole called a thyrsus, held by the adjoining constellation of Centaurus, the centaur. Consequently, the constellations of the Centaur and the animal were often regarded as a combined figure, although Ptolemy listed them as separate constellations in the *Almagest*.

The Babylonians knew this constellation as UR.IDIM, meaning 'wild dog' or 'wolf'. Eratosthenes said that the Centaur was holding the animal towards the altar (the constellation Ara) as though about to sacrifice it. Hyginus referred to the animal as simply 'a victim', while Germanicus Caesar said that the Centaur was either carrying game from the woods, or was bringing gifts to the altar. The identification of this constellation with a wolf seems to have started in Renaissance times, evidently reviving the Babylonian identity.

One is tempted to recall the story of Lycaon, king of the Arcadians, who served Zeus with the flesh of the god's own son and was punished by being

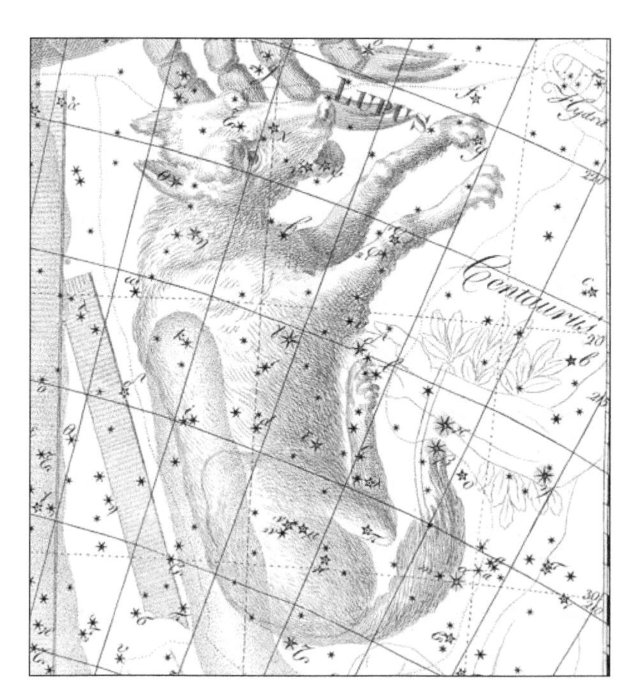

Lupus, the wolf, is visualized as being impaled on a pole (a thyrsus) held by Centaurus, off the chart at right. Centaurus is holding Lupus out towards Ara, the altar (off picture at left), as though about to sacrifice it. This illustration is from Chart XV of the *Uranographia* of Johann Bode (1801). (Deutsches Museum)

turned into a wolf (see Boötes, page 53). But that story has no connection with this constellation, which seems to have been overlooked by the mythologists. The fact that it was imported from the Babylonians probably explains why the Greeks had no myths for it. None of the stars of Lupus is named.

Lynx
The lynx

Genitive: Lyncis
Abbreviation: Lyn
Size ranking: 28th
Origin: The seven constellations of Johannes Hevelius

Johannes Hevelius, the Polish astronomer who introduced this constellation in 1687, continued to measure star positions with the naked eye long after other astronomers had adopted telescopic sights. The French astronomer Pierre Gassendi wrote that Hevelius had the 'eyes of a lynx' and this constellation can be seen as an attempt to demonstrate that. Indeed, Hevelius wrote in his *Prodromus Astronomiae* that anyone who wanted to observe it would need the eyesight of a lynx ('oculos habeat Lynceos'), although he undoubtedly exaggerated the faintness of the 19 stars he catalogued in it, typically by a full magnitude.

Lynx fills a blank area of sky between Ursa Major and Auriga that is surprisingly large – greater in area than Gemini, for example – but apart from one

Lynx shown on Chart V of the *Uranographia* of Johann Bode (1801). (Deutsches Museum)

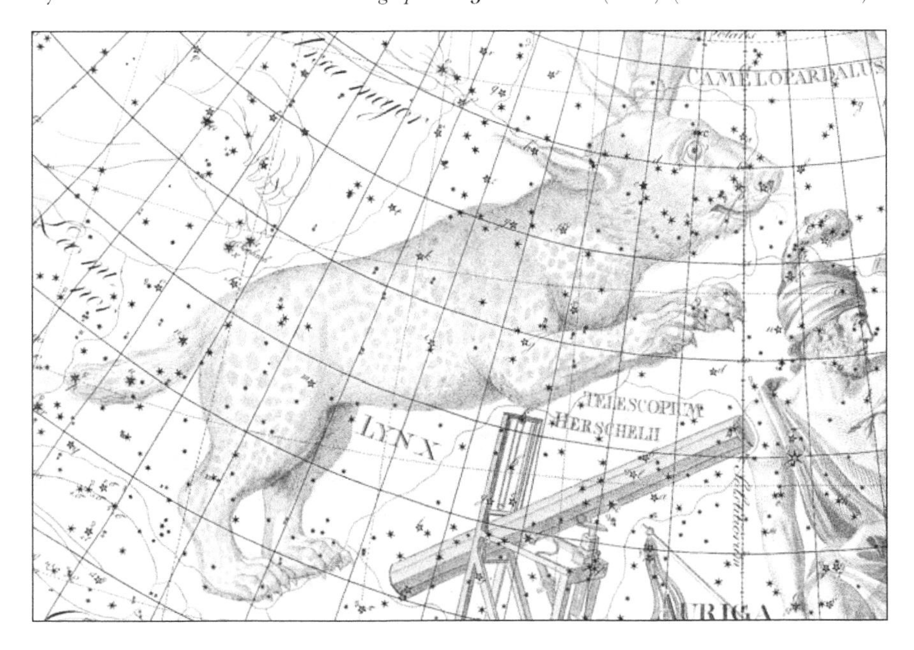

third-magnitude star (Alpha Lyncis) it contains no stars brighter than fourth magnitude. Several of its stars had been listed by Ptolemy in his *Almagest* as lying among what he termed the 'unformed' stars outside Ursa Major. Earlier in the 17th century the Dutch astronomer Petrus Plancius had made these stars part of his new constellation Jordanus, the river Jordan (page 195), but it was Hevelius's alternative creation that endured. On his star atlas *Firmamentum Sobiescianum* Hevelius called the constellation Lynx, while in the accompanying star catalogue it is listed as 'Lynx, sive Tigris' (Lynx or Tiger). However, the illustration he presented does not look much like either animal.

It is not known whether Hevelius had in mind the mythological character Lynceus who enjoyed the keenest eyesight in the world – he was even credited with the ability to see things underground. Lynceus and his twin brother Idas sailed with the Argonauts. The pair came to grief when they fell out with those other mythical twins, Castor and Polydeuces (see Gemini, page 99).

Alpha Lyncis, magnitude 3.1, is the brightest star in the constellation, and the only one with a Greek letter; the letter was assigned by the English astronomer Francis Baily in his *British Association Catalogue* of 1845.

Lyra
The lyre

Genitive: Lyrae
Abbreviation: Lyr
Size ranking: 52nd
Origin: One of the 48 Greek constellations listed by Ptolemy in the *Almagest*
Greek name: Λύρα (Lyra)

A compact but prominent constellation, marked by the fifth-brightest star in the sky, Vega. Mythologically, Lyra (Λύρα in Greek) was the lyre of the great musician Orpheus, whose venture into the Underworld is one of the most famous of Greek stories. It was the first lyre ever made, having been invented by Hermes, the son of Zeus and Maia (one of the Pleiades). Hermes fashioned the lyre from the shell of a tortoise that he found browsing outside his cave on Mount Cyllene in Arcadia. Hermes cleaned out the shell, pierced its rim and tied across it seven strings of cow gut, the same as the number of the Pleiades. He also invented the plectrum with which to play the instrument.

The lyre got Hermes out of trouble after a youthful exploit in which he stole some of Apollo's cattle. Apollo angrily came to demand their return, but when he heard the beautiful music of the lyre he let Hermes keep the cattle and took the lyre in exchange. Eratosthenes says that Apollo later gave the lyre to Orpheus to accompany his songs.

Orpheus was the greatest musician of his age, able to charm rocks and streams with the magic of his songs. He was even reputed to have attracted rows of oak trees down to the coast of Thrace with the music of his lyre. Orpheus

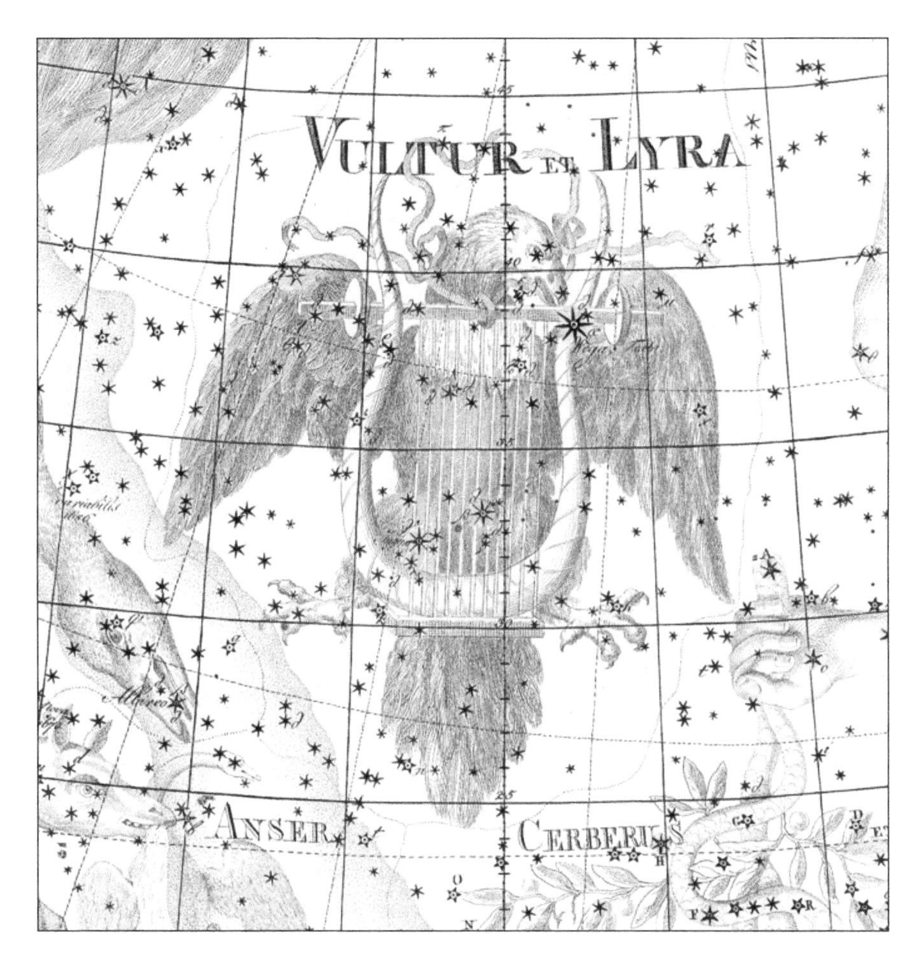

Lyra was frequently visualized as an eagle or vulture as well as a lyre; both are shown on this engraving from Chart VIII of Johann Bode's *Uranographia* (1801). Near the tip of the vulture's beak is the bright star Vega. (Deutsches Museum)

joined the expedition of Jason and the Argonauts in search of the golden fleece. When the Argonauts heard the tempting song of the Sirens, sea nymphs who had lured generations of sailors to destruction, Orpheus sang a counter melody that drowned the Sirens' voices.

Orpheus and Eurydice

Later, Orpheus married the nymph Eurydice. One day, Eurydice was spied by Aristaeus, a son of Apollo, who attacked her in a fit of passion. Fleeing from him, she stepped on a snake and died from its poisonous bite. Orpheus was heartbroken; unable to live without his young wife, Orpheus descended into the Underworld to plead for her release. Such a request was unprecedented. But the sound of his music charmed even the cold heart of Hades, god of the Underworld, who finally agreed to let Eurydice accompany Orpheus back to the land of the living on one solemn condition: Orpheus must not at any stage look behind him until the couple were safely back in daylight.

Orpheus readily accepted, and led Eurydice through the dark passage that led to the upper world, strumming his lyre to guide her. It was an unnerving feeling to be followed by a ghost. He could never be quite sure that his beloved was following, but he dared not look back. Eventually, as they approached the surface, his nerve gave out. He turned around to confirm that Eurydice was still there – and at that moment she slipped back into the depths of the Underworld, out of his grasp for ever.

Orpheus was inconsolable. Thereafter he wandered the countryside, plaintively playing his lyre. Many women offered themselves to the great musician in marriage, but he preferred the company of young boys.

The death of Orpheus

There are two accounts of the death of Orpheus. One version, told by Ovid in his *Metamorphoses*, says that the local women, offended at being rejected by Orpheus, ganged up on him as he sat singing one day. They began to throw rocks and spears at him. At first his music charmed the weapons so that they fell harmlessly at his feet, but the women raised such a din that they eventually drowned the magic music and the missiles found their mark.

Eratosthenes, on the other hand, says that Orpheus incurred the wrath of the god Dionysus by not making sacrifices to him. Orpheus regarded Apollo, the Sun god, as the supreme deity and would often sit on the summit of Mount Pangaeum awaiting dawn so that he could be the first to salute the Sun with his melodies. In retribution for this snub, Dionysus sent his manic followers to tear Orpheus limb from limb. Either way, Orpheus finally joined his beloved Eurydice in the Underworld, while the muses put the lyre among the stars with the approval of Zeus, their father.

Vega and other stars

Ptolemy knew the constellation's brightest star simply as Λύρα (Lyra), the same name as the constellation. The name we use for this star today, Vega, comes from the Arabic words *al-nasr al-waqi'* which can mean either 'the swooping eagle' or 'vulture', for the Arabs saw both an eagle and a vulture here. The constellation was often depicted on star maps as a bird positioned behind a lyre, as in the illustration on the previous page. It seems that the Arabs visualized Vega and its two nearby stars Epsilon and Zeta Lyrae as an eagle with folded wings, swooping down in its prey, whereas in the nearby constellation Aquila the star Altair and its two attendant stars gave the impression of a flying eagle with wings outstretched. Vega is 25 light years away.

Beta Lyrae is called Sheliak, a name that comes from the Arabic for 'harp', in reference to the constellation as a whole. Beta Lyrae is a celebrated variable star. Gamma Lyrae is called Sulafat, from the Arabic meaning 'the tortoise', after the animal from whose shell Hermes made the lyre. Between Beta and Gamma Lyrae lies the Ring Nebula, often pictured in astronomy books; it is a shell of gas like a celestial smoke ring, thrown off by a dying star.

Mensa

The table mountain

Genitive: Mensae
Abbreviation: Men
Size ranking: 75th
Origin: The 14 southern constellations of Nicolas Louis de Lacaille

A small, faint constellation of the far southern sky invented by the French astronomer Nicolas Louis de Lacaille to commemorate Table Mountain near Cape Town, South Africa, from where he catalogued the southern stars in 1751–52. Lacaille originally gave it the French name Montagne de la Table on the first version of his planisphere published in 1756 but this was Latinized to Mons Mensae on the second edition of 1763. In 1844 the English astronomer John Herschel proposed shortening it to Mensa. Francis Baily adopted this suggestion in his *British Association Catalogue* of 1845, and it has been known as Mensa ever since.

Mensa contains part of the Large Magellanic Cloud, a neighbour galaxy to our Milky Way, which gives Mensa the appearance of being capped by a white cloud, like the so-called 'table-cloth' cloud sometimes seen over the real Table Mountain 'at the approach of a violent south-easterly wind', as Lacaille put it. Mensa's brightest stars are of only fifth magnitude.

Mensa, under its original name Mons Mensae, on Chart XX of the *Uranographia* of Johann Bode (1801). Nubecula Major, here seen below left of it, is the Large Magellanic Cloud, reminiscent of the cloud that caps the real Table Mountain. (Deutsches Museum)

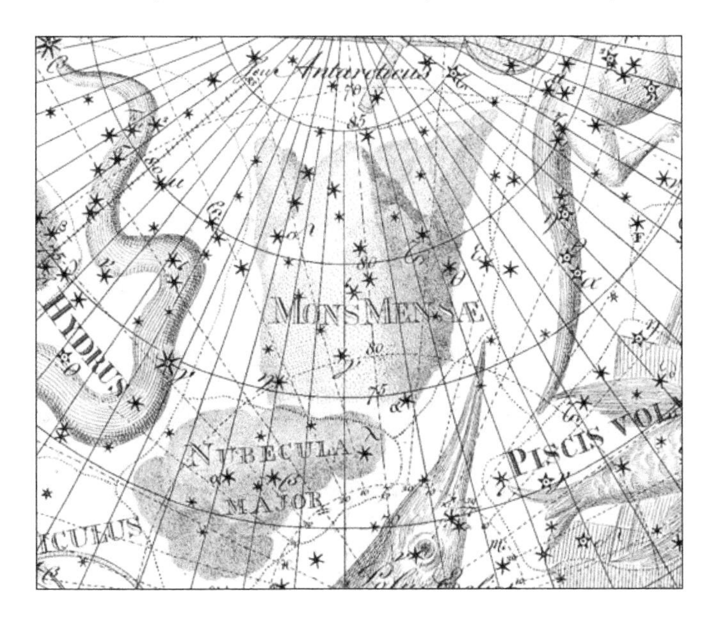

Microscopium

The microscope

Genitive: Microscopii
Abbreviation: Mic
Size ranking: 66th
Origin: The 14 southern constellations of Nicolas Louis de Lacaille

One of the southern constellations representing scientific instruments that were invented in 1751–52 by the French astronomer Nicolas Louis de Lacaille. In this case the instrument concerned is an early form of compound microscope, i.e. one that uses more than one lens.

Lacaille first showed the constellation on his map of 1756 under the name le Microscope but Latinized this to Microscopium on the second edition of the chart published in 1763. He described it as consisting of 'a tube above a square box', although Johann Bode added a slide carrier containing specimens when he depicted it in his *Uranographia* atlas of 1801 (see below).

Microscopium lies beneath the forelegs of Capricornus in an area of sky containing stars of only fifth magnitude and fainter. It incorporates six 'unformed' stars that Ptolemy had listed as lying outside Piscis Austrinus. One of these is the constellation's brightest star, Gamma Microscopii, magnitude 4.7, seen on the side of the tube in the chart below. The eyepiece end of the microscope is marked by 6th-magnitude Delta Microscopii. The only remarkable thing about Microscopium is that anyone could imagine a separate constellation here.

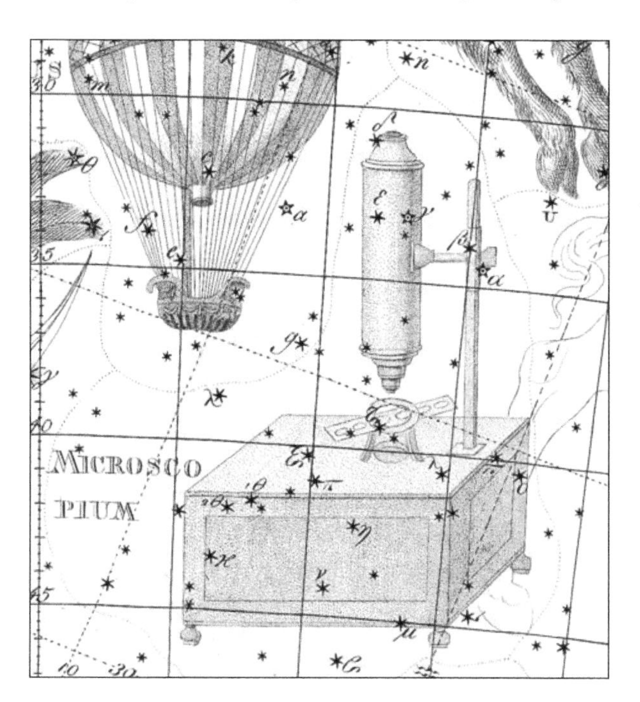

Microscopium shown ready to examine a specimen on a glass slide as seen on Chart XVI of the *Uranographia* of Johann Bode (1801). Next to it on this chart lies the obsolete constellation Globus Aerostaticus, the hot-air balloon (see page 193). Above it are the forelegs of Capricornus. (Deutsches Museum)

Monoceros
The unicorn

Genitive: Monocerotis
Abbreviation: Mon
Size ranking: 35th
Origin: Petrus Plancius

The mythical single-horned beast, the unicorn, is represented by this constellation which was unknown to the ancient Greeks. Monoceros was first depicted in 1612 under the name Monoceros Unicornis on a globe by the Dutch theologian and cartographer Petrus Plancius. This was the same globe on which Camelopardalis, another of his inventions, first appeared.

In 1624 the German astronomer Jacob Bartsch depicted it under the name Unicornu (*sic*) on a star chart in his book *Usus Astronomicus Planisphaerii Stellati* and as a result he was sometimes wrongly credited with its invention. In his book, Bartsch pointed to several passages in the Bible that supposedly mention unicorns, although these are now regarded as mistranslations. It is not clear whether Plancius introduced the constellation because of these Biblical references, but the unicorn has long been regarded as a Christian symbol of purity. Perhaps Plancius had in mind the Hunt of the Unicorn tapestries woven in the

Monoceros prances between Orion's dogs Canis Major (below it) and Canis Minor (above) on the *Atlas Coelestis* of John Flamsteed (1729). (University of Michigan Library)

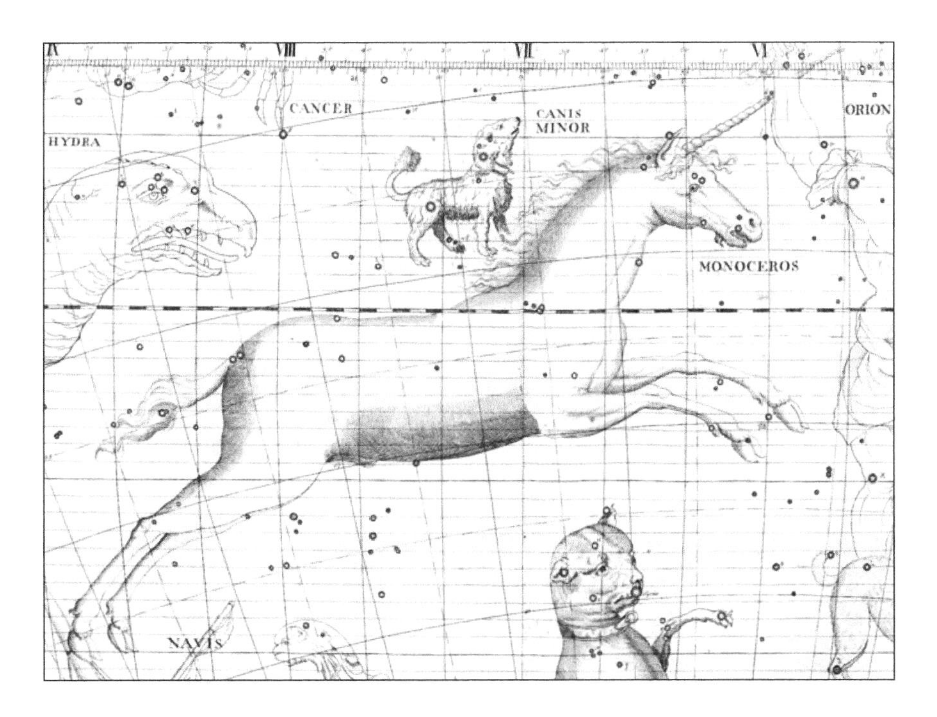

southern Netherlands around a century earlier. The Polish astronomer Johannes Hevelius adopted Monoceros in his influential star atlas and catalogue published in 1690 which ensured its acceptance by other astronomers.

Monoceros fills a large area between Hydra and Orion where there was no Greek constellation. It is not prominent (its brightest stars are of fourth magnitude) but it lies in the Milky Way and contains a host of fascinating objects, most notably the Rosette Nebula, a wreath-shaped mass of glowing gas with embedded stars. There are no legends associated with the constellation, as it is a modern figure, and none of its stars is named.

Musca
The fly

Genitive: Muscae
Abbreviation: Mus
Size ranking: 77th
Origin: The 12 southern constellations of Keyser and de Houtman

A small constellation to the south of Crux, the southern cross. Musca was one of the 12 southern constellations introduced at the end of the 16th century by Pieter Dirkszoon Keyser and Frederick de Houtman from the stars they observed during the first Dutch expeditions to the East Indies. It was first depicted by their fellow Dutchman Petrus Plancius on his globe of 1598, but

Musca, shown under its sometime alternative name of Apis, the bee (but looking more like a wasp), on Chart XX of the *Uranographia* of Johann Bode (1801). (Deutsches Museum)

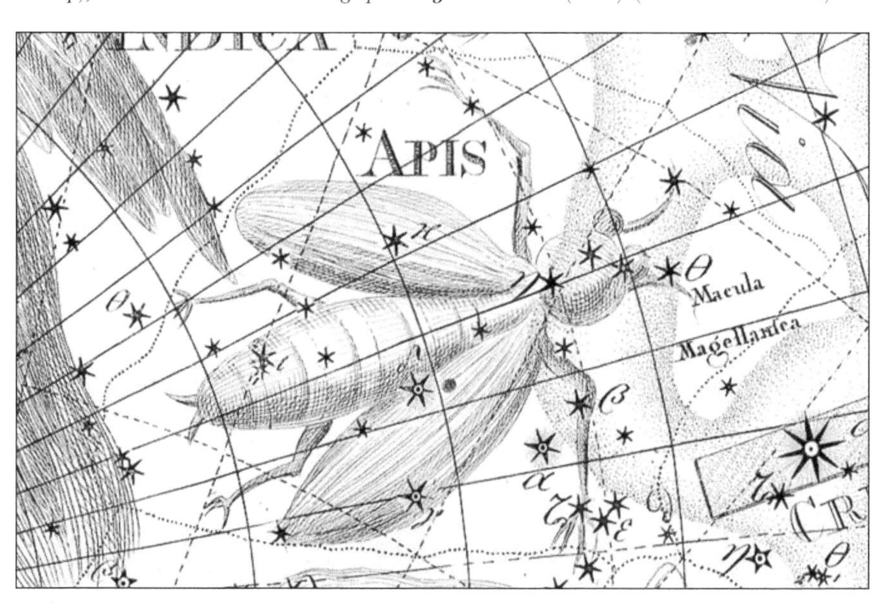

for some reason he left it unnamed. In de Houtman's catalogue of 1603, completed after Keyser's death, it is called De Vlieghe, Dutch for fly.

Johann Bayer, also in 1603, showed the insect on his plate of the 12 new southern constellations in *Uranometria* but called it Apis, the bee, an alternative title which was widely used for two centuries. The Dutch historian Elly Dekker believes that this alternative identification arose because Bayer copied his southern constellations from globes produced by Jodocus Hondius (1563–1612) in 1600 and 1601, on which the figure was left unnamed. Not knowing what it was meant to depict, Bayer wrongly identified it as a bee (apis), not a fly (musca).

The first known use of the Latin name Musca for this constellation was in 1602 on a globe by Willem Janszoon Blaeu (1571–1638), another Dutch cartographer and rival to Plancius. Plancius himself did not adopt a name for the constellation until 1612, when he called it Muia, the Greek for fly, on a globe produced that year. For a time it was known as Musca Australis, when there was also a northern fly, Musca Borealis, in the sky (page 198).

The constellation's brightest star, Alpha Muscae, is of third magnitude. None of its stars is named, and there are no legends about the fly.

Norma

The set square

Genitive: Normae
Abbreviation: Nor
Size ranking: 74th
Origin: The 14 southern constellations of Nicolas Louis de Lacaille

Norma is one of the constellations introduced by the French astronomer Nicolas Louis de Lacaille following his mapping of the southern skies in 1751–52. It consists of faint stars between Ara and Lupus that were not catalogued by Ptolemy. On his 1756 planisphere Lacaille called it l'Equerre et la Regle, the square and rule, depicting it as a draughtsman's set-square and ruler.

Lacaille placed it next to another of his inventions, the compasses (which he called le Compas, now known as Circinus), and the southern triangle (Triangulum Australe); this latter figure was an earlier invention of the Dutch navigators Keyser and de Houtman which Lacaille visualized as a builder's level, thereby creating a trio of surveying and building instruments. On the 1763 edition of the planisphere Lacaille Latinized and shortened the name of the constellation to Norma; others, though, preferred the fuller name Norma et Regula, as did Johann Bode on his atlas of 1801 (see illustration on the following page). The brightest stars of Norma are of only fourth magnitude and none have names.

In his widely quoted book *Star Names, Their Lore and Meaning* the historian R. H. Allen called this constellation 'the Level and Square'. Allen said that the French edition of Flamsteed's star atlas (i.e. the *Atlas Céleste* of Jean Fortin) showed it as Niveau, the level, but he is wrong – the alternative name 'level' was

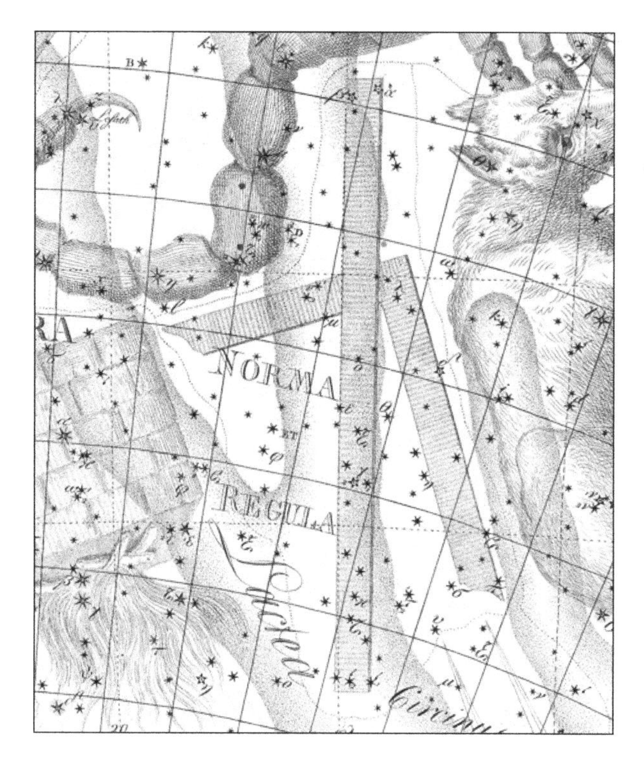

Norma, shown under its old name Norma et Regula on Chart XV of the *Uranographia* of Johann Bode (1801). It lies in the Milky Way between Ara, the altar, to its left, and Lupus, the wolf, on the right. Above it is the curling tail of Scorpius, the scorpion. 'Rule and Square' is a traditional English pub name. (Deutsches Museum)

actually applied to the southern triangle, Triangulum Australe. Allen seemingly misread the French map and transferred the name to the wrong constellation, in a reprise of the way in which star names were misapplied through misreadings of maps down the ages. Allen's error has caused confusion ever since.

Because of changes in the constellation's boundaries since Lacaille's time, Norma no longer has stars labelled Alpha or Beta. The stars that Lacaille designated Alpha and Beta Normae are now part of Scorpius, where they are known as N and H Scorpii respectively.

Octans

The octant

Genitive: Octantis
Abbreviation: Oct
Size ranking: 50th
Origin: The 14 southern constellations of Nicolas Louis de Lacaille

Octans was one of 14 new southern constellations introduced in the 1750s by the French astronomer Nicolas Louis de Lacaille. It represents a navigational instrument known as a reflecting octant, invented in 1730 by the Englishman John Hadley (1682–1744). Lacaille originally named it l'Octans de Reflexion on his chart published in 1756, but changed this simply to Octans on the second edition in 1763. When constructing the constellation Lacaille annexed several stars which had previously been regarded as part of Hydrus, the lesser water snake, one of the constellations formed at the end of the 16th century by Dutch navigators.

An octant consists of an arc of 45°, i.e. an eighth of a circle, hence the name. The navigator sighted the horizon through a telescope and adjusted a movable arm until the reflected image of the Sun or a star overlay the direct view of the horizon. In later designs the arc was extended from one-eighth of a circle to one-sixth and the instrument became the modern sextant.

Fittingly enough for a navigational instrument, Octans encompasses the south celestial pole, but despite this privileged position it contains little of note, consisting of no stars brighter than fourth magnitude. There is, unfortunately, no southern equivalent of the bright northern pole star, Polaris. The nearest naked-eye star to the south celestial pole is Sigma Octantis, a degree away from the pole, although at magnitude 5.4 it is far from prominent.

Octans is another example of a constellation in which the star labelled Alpha is not the brightest. In this case, the brightest star is Nu Octantis, magnitude 3.7. Alpha Octantis, magnitude 5.2, has the distinction of being the faintest star labelled Alpha in any constellation. Its closest rival, Alpha Mensae, is about 0.1 of a magnitude brighter.

Octans encompasses the south celestial pole, as shown on Chart XX of Johann Bode's *Uranographia* star atlas, where it was called Octans Nautica. The octant was the forerunner of the modern sextant. (Deutsches Museum)

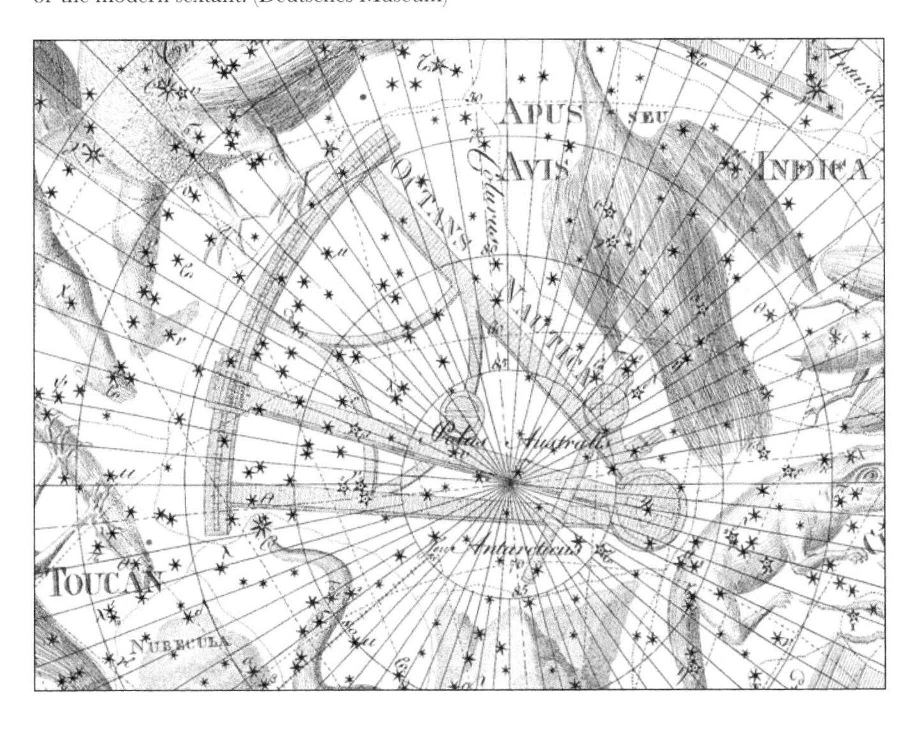

Ophiuchus
The serpent holder

Genitive: Ophiuchi
Abbreviation: Oph
Size ranking: 11th
Origin: One of the 48 Greek constellations listed by Ptolemy in the *Almagest*
Greek name: Ὀφιοῦχος (Ophiouchos)

Ophiuchus (pronounced off-ee-YOO-cuss) represents a man with a snake coiled around his waist. He holds the head of the snake in his left hand and its tail in his right hand. The snake is represented by a separate constellation, Serpens. The Greek spelling of the name was Ὀφιοῦχος (Ophiouchos).

The Greeks identified him as Asclepius, the god of medicine (Aesculapius in Latin). Asclepius was the son of Apollo and Coronis (although some say that his mother was Arsinoë). The story goes that Coronis two-timed Apollo by sleeping with a mortal, Ischys, while she was pregnant by Apollo. A crow brought Apollo the unwelcome news, but instead of the expected reward the crow, which until then had been snow-white, was cursed by Apollo and turned black.

In a rage of jealousy, Apollo shot Coronis with an arrow. Rather than see his child perish with her, Apollo snatched the unborn baby from its mother's womb as the flames of the funeral pyre engulfed her, and took the infant to Chiron, the wise centaur (represented in the sky by the constellation Centaurus). Chiron raised Asclepius as his own son, teaching him the arts of healing and hunting. Asclepius became so skilled in medicine that not only could he save lives, he could also raise the dead.

Asclepius and the snake

On one occasion in Crete, Glaucus, the young son of King Minos, fell into jar of honey while playing and drowned. As Asclepius contemplated the body of Glaucus, a snake slithered towards it. He killed the snake with his staff; then another snake came along with a herb in its mouth and placed it on the body of the dead snake, which magically returned to life. Asclepius took the same herb and laid it on the body of Glaucus, who too was magically resurrected. (Robert Graves suggests that the herb was mistletoe, which the ancients thought had great regenerative properties, but perhaps it was actually willow bark, the source of salicylic acid, the active ingredient in aspirin.)

Because of this incident, says Hyginus, Ophiuchus is shown in the sky holding a snake, which became the symbol of healing from the fact that snakes shed their skin every year and are thus seemingly reborn. Others, though, say that Asclepius received from the goddess Athene the blood of Medusa the Gorgon. The blood that flowed from the veins on her left side was a poison, but the blood from the right side could raise the dead.

Someone else supposedly resurrected by Asclepius was Hippolytus, son of Theseus, who died when he was thrown from his chariot; some identify him

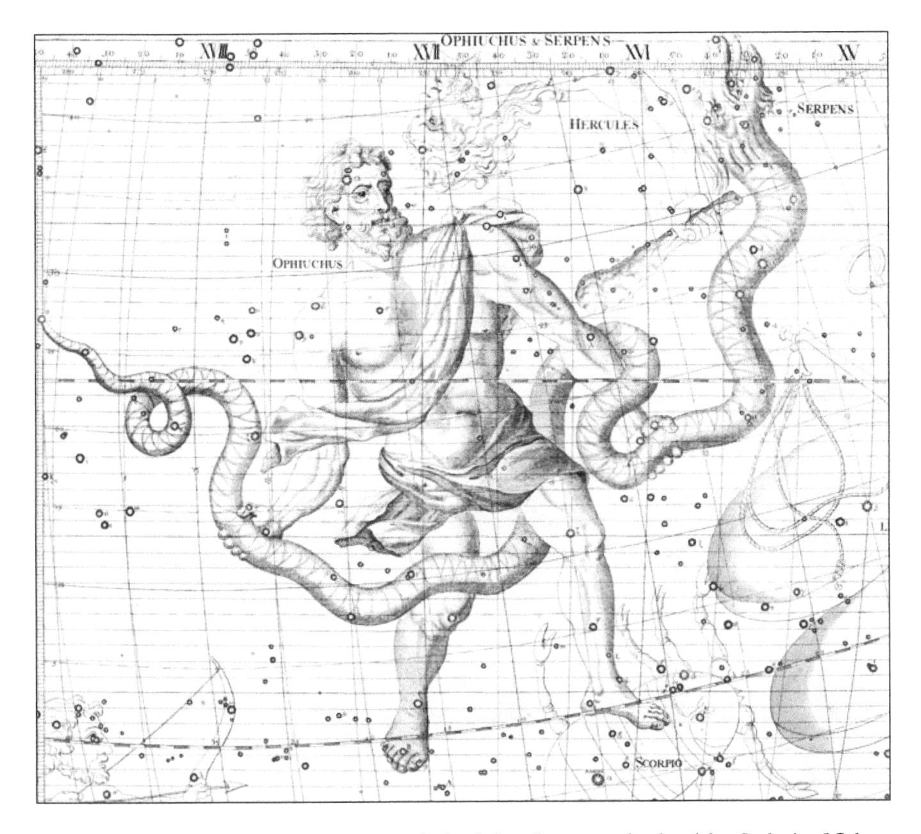

Ophiuchus holds a huge snake, Serpens, in both hands as seen in the *Atlas Coelestis* of John Flamsteed (1729). His left foot rests on Scorpius, the scorpion. The dashed line crossing the feet of Ophiuchus is the Sun's annual path, the ecliptic. (University of Michigan Library)

with the constellation Auriga, the charioteer. Reaching for his healing herbs, Asclepius touched the youth's chest three times, uttering healing words, and Hippolytus raised his head.

Hades, god of the Underworld, began to realize that the flow of dead souls into his domain would soon dry up if this technique became widely known. He complained to his brother god Zeus who struck down Asclepius with a thunderbolt. Apollo was outraged at this harsh treatment of his son and retaliated by killing the three Cyclopes who forged Zeus' thunderbolts. To mollify Apollo, Zeus made Asclepius immortal (in the circumstances he could hardly bring him back to life again) and set him among the stars as the constellation Ophiuchus.

Stars of Ophiuchus

The head of Ophiuchus is marked by its brightest star, second-magnitude Alpha Ophiuchi, called Rasalhague from the Arabic meaning 'the head of the serpent collector'. Beta Ophiuchi is called Cebalrai from the Arabic for 'the shepherd's dog'; the Arabs visualized a shepherd (the star Alpha Ophiuchi) along with his dog and some sheep in this area. Ptolemy in the *Almagest* located Beta and

Gamma Ophiuchi in the serpent holder's right shoulder; the left shoulder is marked by Iota and Kappa Ophiuchi.

Delta and Epsilon Ophiuchi are called Yed Prior and Yed Posterior. These are compound names, formed from the Arabic *al-yad*, meaning 'hand', with the Latin words prior and posterior added to give names meaning the 'leading' and 'following' parts of the hand. The hand in question is the left one; the right hand, according to Ptolemy, was marked by the stars we know as Nu and Tau Ophiuchi, but these have no proper names.

Zeta and Eta Ophiuchi are his left and right knees, while Rho and Theta Ophiuchi are in his feet. Scorpius, the scorpion, lies beneath his feet. Aratus said that Ophiuchus 'tramples' the scorpion with both feet, but in reality it is only the left foot that stands on the scorpion; the right foot remains well clear of it.

Between the right shoulder of Ophiuchus and the tail of the serpent lies a V-shaped group of five stars that Ptolemy regarded as being outside the main figure of Ophiuchus. These stars were later incorporated into the short-lived constellation Taurus Poniatovii (page 204). Barnard's Star, the second-closest star to the Sun at 5.9 light years away, lies in this same area, near 66 Ophiuchi.

Ophiuchus was the site of the last supernova seen in our Galaxy. This appeared in 1604 near Xi Ophiuchi and reached an estimated maximum magnitude of −3. It is known as Kepler's Star after Johannes Kepler who described it in a book called *De stella nova* (1606).

Ophiuchus and the zodiac

Although Ophiuchus is not one of the official 12 constellations of the zodiac, the Sun passes through its southern regions in the first half of December. The Sun's path, the ecliptic, is the black-and-white line crossing the feet of Ophiuchus in the chart on the previous page. According to the modern constellation boundaries, the Sun spends longer in Ophiuchus than it does in adjoining Scorpius. Hence Ophiuchus is sometimes referred to as the 13th sign of the zodiac.

Orion

The hunter

Genitive: Orionis
Abbreviation: Ori
Size ranking: 26th
Origin: One of the 48 Greek constellations listed by Ptolemy in the *Almagest*
Greek name: 'Ωρίων (Orion)

Orion ('Ωρίων in Greek) is the most splendid of constellations, befitting a character who was in legend the tallest and most handsome of men. His right shoulder and left foot are marked by the brilliant stars Betelgeuse and Rigel, with a distinctive line of three stars forming his belt. 'No other constellation more accurately represents the figure of a man', says Germanicus Caesar. It is

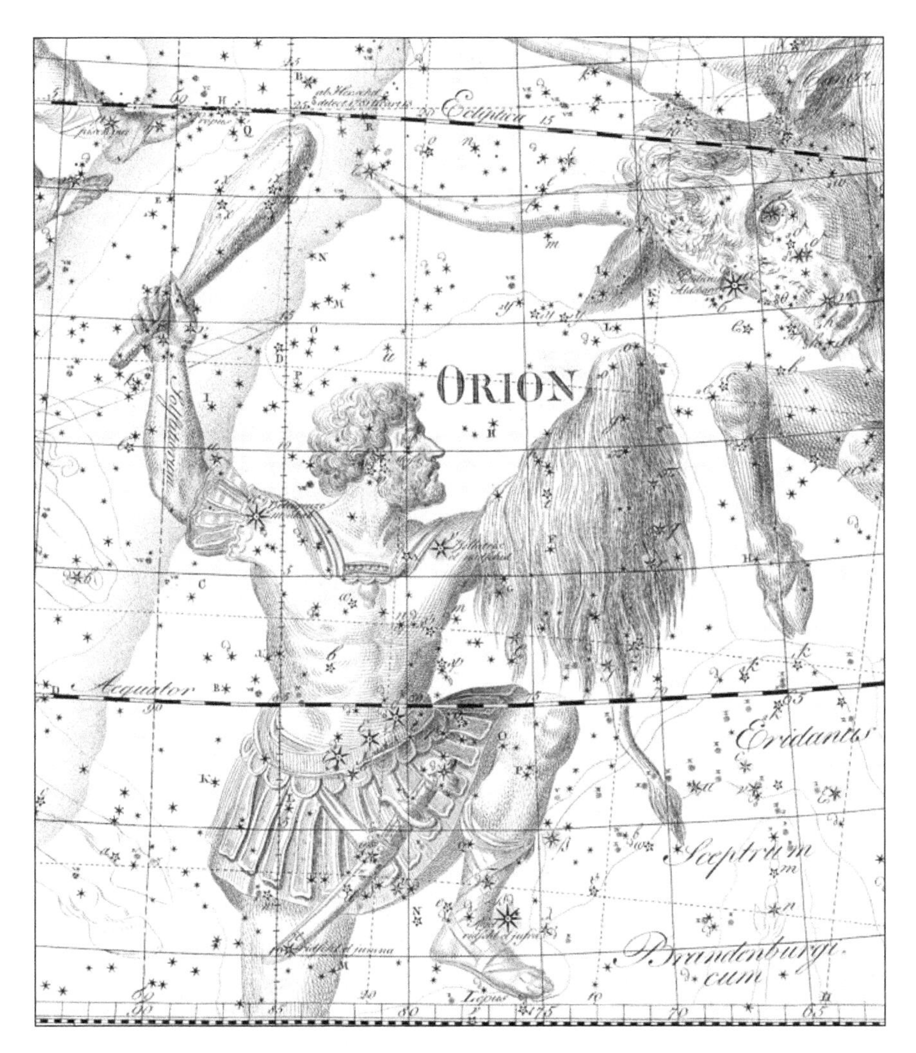

Orion raises his club and shield against the charging Taurus on Chart XII from Johann Bode's *Uranographia* (1801). Orion's right shoulder is marked by the bright star Betelgeuse, and his left foot by Rigel. A line of three stars forms his belt from which hangs his sword, marked by the glowing gas cloud called the Orion Nebula. (Deutsches Museum)

also one of the most ancient constellations, being among the few star groups known to the earliest Greek writers such as Homer and Hesiod *c.*700 BC. Even in the space age, Orion remains one of the few star patterns that non-astronomers can recognize.

Manilius called it 'golden Orion' and 'the mightiest of constellations', and exaggerated its brilliance by saying that, when Orion rises, 'night feigns the brightness of day and folds its dusky wings'. Manilius described Orion as 'stretching his arms over a vast expanse of sky and rising to the stars with no less huge a stride'. In fact, Orion is not an exceptionally large constellation, ranking only 26th in size (smaller, for instance, than Perseus according to the modern constellation boundaries), but the brilliance of its stars gives it the illusion of being much larger.

In the sky, Orion is depicted facing the snorting charge of neighbouring Taurus the bull, yet the myth of Orion makes no reference to such a combat. However, the constellation originated with the Sumerians, who saw in it their great hero Gilgamesh fighting the Bull of Heaven. The Sumerian name for Orion was URU AN-NA, meaning light of heaven. Taurus was GUD AN-NA, bull of heaven.

Gilgamesh was the Sumerian equivalent of Heracles (Hercules in Latin), which brings us to another puzzle. Being the greatest hero of Greek mythology, Heracles deserves a magnificent constellation such as this one, but in fact is consigned to a much more obscure area of sky. So is Orion really Heracles in another guise? It might seem so, for one of the labours of Heracles was to catch the Cretan bull, which would fit the Orion–Taurus conflict in the sky. Ptolemy described him with club and lion's pelt, both familiar attributes of Heracles, and he is shown this way on old star maps. Despite these parallels, no mythologist hints at a connection between this constellation and Heracles.

Tales of Orion

According to Greek myth, Orion was the son of Poseidon the sea god and Euryale, daughter of King Minos of Crete. Poseidon gave Orion the power to walk on water. Homer in the *Odyssey* describes Orion as a giant hunter, armed with an unbreakable club of solid bronze. In the sky, the hunter's dogs (the constellations Canis Major and Canis Minor) follow at his heels, in pursuit of the hare (the constellation Lepus).

On the island of Chios, Orion wooed Merope, daughter of King Oenopion, apparently without much success, for one night while fortified with wine he tried to ravish her. In punishment, Oenopion put out Orion's eyes and banished him from the island. Orion headed north to the island of Lemnos where Hephaestus had his forge. Hephaestus took pity on the blind Orion and offered one of his assistants, Cedalion, to act as his eyes. Hoisting the youth on his shoulders, Orion headed east towards the sunrise, which an oracle had told him would restore his sight. As the Sun's healing rays fell on his sightless eyes at dawn, Orion's vision was miraculously restored.

Orion is linked in a stellar myth with the Pleiades star cluster in Taurus. The Pleiades were seven sisters, daughters of Atlas and Pleione. As the story is usually told, Orion fell in love with the Pleiades and pursued them with amorous intent. But according to Hyginus, it was actually their mother Pleione he was after. Zeus snatched the group up and placed them among the stars, where Orion still pursues them across the sky each night.

There is a strange and persistent story about the birth of Orion, designed to account for the early version of his name, Urion (even closer to the Sumerian original URU AN-NA). According to this story, there lived in Thebes an old farmer named Hyrieus. One day he offered hospitality to three passing strangers, who happened to be the gods Zeus, Neptune, and Hermes. After they had eaten, the visitors asked Hyrieus if he had any wishes. The old man confessed that he would have liked a son, and the three gods promised to fulfil his wish. Standing together around the hide of the ox they had just consumed, the gods urinated on it and told Hyrieus to bury the hide. From it in due course was born a boy whom Hyrieus named Urion after the mode of his conception.

Death of Orion

Stories of the death of Orion are numerous and conflicting. Astronomical mythographers such as Aratus, Eratosthenes, and Hyginus were agreed that a scorpion was involved. In one version, told by Eratosthenes and Hyginus, Orion boasted that he was the greatest of hunters. He declared to Artemis, the goddess of hunting, and Leto, her mother, that he could kill any beast on Earth. The Earth shuddered indignantly and from a crack in the ground emerged a scorpion which stung the presumptuous giant to death.

Aratus, though, says that Orion attempted to ravish the virgin Artemis, and it was she who caused the Earth to open, bringing forth the scorpion. Ovid has still another account; he says that Orion was killed trying to save Leto from the scorpion. Even the location varies. Eratosthenes and Hyginus say that Orion's death happened in Crete, but Aratus places it in Chios.

In both versions, the outcome was that Orion and the scorpion (the constellation Scorpius) were placed on opposite sides of the sky, so that as Scorpius rises in the east, Orion flees below the western horizon. 'Wretched Orion still fears being wounded by the poisonous sting of the scorpion', noted Germanicus Caesar.

A very different story, also recounted by Hyginus, is that Artemis loved Orion and was seriously considering giving up her vows of chastity to marry him. As the greatest male and female hunters they would have made a formidable couple. But Apollo, twin brother of Artemis, was against the match. One day, while Orion was swimming, Apollo challenged Artemis to demonstrate her skill at archery by hitting a small black object that he pointed out bobbing among the waves. Artemis pierced it with one shot – and was horrified to find that she had killed Orion. Grieving, she placed him among the constellations.

Bright stars in Orion

Orion is one of several constellations in which the star labelled Alpha is not the brightest. The brightest star in Orion is actually Beta Orionis, called Rigel from the Arabic *rijl* meaning 'foot' from Ptolemy's description of it as 'the bright star in the left foot'. Ptolemy also said it was shared with the river Eridanus, and some old charts depict it in this dual role. Rigel is a brilliant blue-white supergiant about 860 light years away.

Alpha Orionis is called Betelgeuse (pronounced BET-ell-juice), one of the most famous yet misunderstood star names. It comes from the Arabic *yad al-jauza*, often wrongly translated as 'armpit of the central one'. In fact, it means 'hand of *al-jauza*'. But who (or what) was *al-jauza*? It was the name given by the Arabs to the constellation figure that they saw in this area, seemingly a female figure encompassing the stars of both Orion and Gemini. The word *al-jauza* apparently comes from the Arabic *jwz* meaning 'middle', so the best translation that modern commentators can offer is that *al-jauza* means something like 'the female one of the middle'. The reference to the 'middle' may be to do with the fact that the constellation lies astride the celestial equator. Ptolemy described it in the *Almagest* as 'the bright, reddish star on the right shoulder'.

The Greeks did not give a name to either Betelgeuse or Rigel, surprisingly for such prominent stars, which is why we know them by their Arabic titles. Betelgeuse is a red supergiant star hundreds of times the diameter of the Sun. It expands and contracts over periods of months and years, changing brightness noticeably in the process. It lies some 500 light years away.

The left shoulder of Orion is marked by Gamma Orionis, known as Bellatrix, a Latin name meaning 'the female warrior'. The star at the hunter's right knee, Kappa Orionis, is called Saiph. This name comes from the Arabic for 'sword', and is clearly misplaced. The three stars of the belt – Zeta, Epsilon, and Delta Orionis – are called Alnitak, Alnilam, and Mintaka. The names Alnitak and Mintaka both come from the Arabic word meaning 'the belt' or 'girdle'. Alnilam comes from the Arabic meaning 'the string of pearls', another reference to the belt of Orion.

Below the belt lies a hazy patch marking the giant's sword or hunting knife. This is the location of the Orion Nebula, one of the most-photographed objects in the sky, a mass of gas some 1,500 light years away from which a cluster of stars is being born. It is visible to the naked eye on clear nights and is easily found with binoculars. The gas of the Nebula shines by the light of the hottest stars that have already formed within. The four brightest of them form a cluster called the Trapezium.

Pavo

The peacock

Genitive: Pavonis
Abbreviation: Pav
Size ranking: 44th
Origin: The 12 southern constellations of Keyser and de Houtman

The peacock is one of the 12 figures introduced into the southern skies at the end of the 16th century by the Dutch navigators Pieter Dirkszoon Keyser and Frederick de Houtman. Pavo probably represents not the common blue, or Indian, peacock commonly seen in parks but its larger, more colourful, and more aggressive cousin, the Java green peacock which Keyser and de Houtman would have encountered in the East Indies. Pavo was first depicted in 1598 on a globe by Petrus Plancius and first appeared in print in 1603 on the *Uranometria* atlas of Johann Bayer.

In mythology the peacock was the sacred bird of Hera, who drove through the air in a chariot drawn by peacocks. How the peacock came to have eyes on

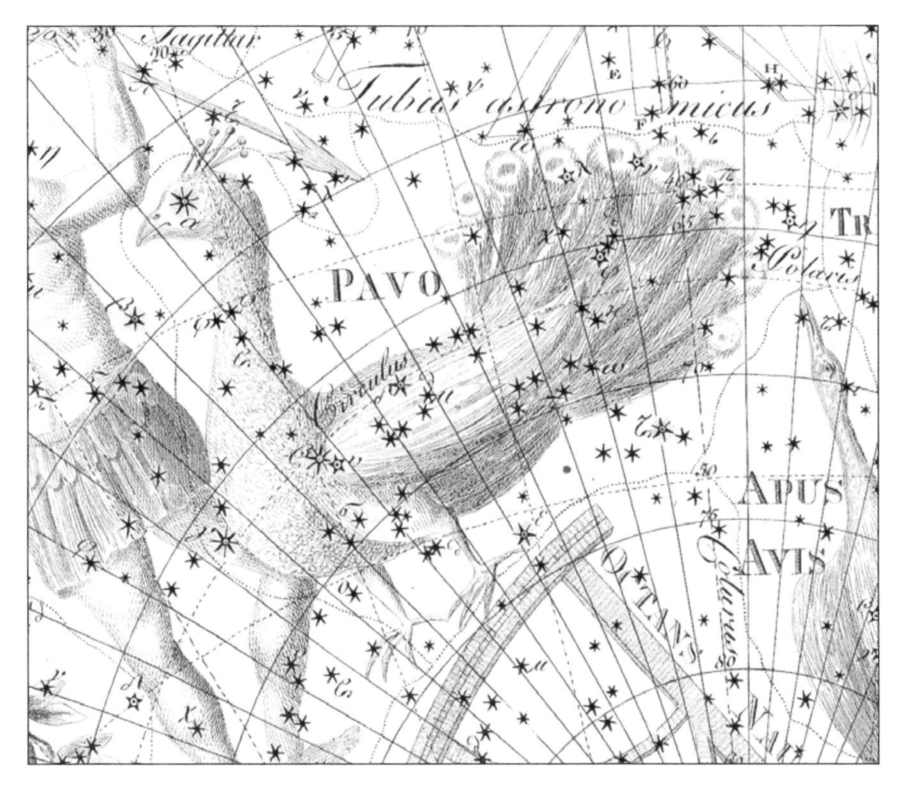

Pavo flourishes a truncated tail on Chart XX of the *Uranographia* of Johann Bode (1801). On Johann Bayer's original representation of 1603 the peacock had a more expansive tail, but this was later trimmed by Lacaille to make room for Telescopium to the north, here called Tubus astronomicus. (Deutsches Museum)

its tail is the subject of a Greek myth that began one day when Zeus turned his illicit love Io into a white cow to disguise her from his wife, Hera, who nearly caught them together. Hera was suspicious and put the heifer under the guardianship of Argus, who tethered the animal to an olive tree. Argus was ideally suited to the task of watchman, since he had 100 eyes, of which only two were resting at a time while the others kept a look out. Wherever Argus stood, he could always keep several of his eyes on Io.

Zeus sent his son Hermes to release Io from her captivity. Hermes swooped down to Earth and spent the day with Argus, telling him stories and playing his reed pipes until, one by one, the eyes of Argus became sleepy and began to close. When Argus was finally asleep, Hermes lopped off his head and released the heifer. Hera placed the eyes of Argus on the tail of the peacock.

The constellation's brightest star, second-magnitude Alpha Pavonis, is called Peacock, a name given in or around 1937 by the UK's Nautical Almanac Office for use in *The Air Almanac*, a navigation guide produced for the Royal Air Force. The RAF specified that all navigation stars should have proper names, so this name was coined for the otherwise unnamed Alpha Pavonis.

Pegasus
The winged horse

Genitive: Pegasi
Abbreviation: Peg
Size ranking: 7th
Origin: One of the 48 Greek constellations listed by Ptolemy in the *Almagest*
Greek name: Ἵππος (Hippos)

Pegasus was the winged horse best known for his association with the Greek hero Bellerophon. The manner of the horse's birth was unusual, to say the least. Its mother was Medusa, the Gorgon, who in her youth was famed for her beauty, particularly her flowing hair. Many suitors approached her, but the one who took her virginity was Poseidon, who is both god of the sea and of horses. Unfortunately, the seduction happened in the temple of Athene. Outraged by having her temple defiled, the goddess Athene changed Medusa into a snake-haired monster whose gaze could turn men to stone.

When Perseus decapitated Medusa, both Pegasus and the warrior Chrysaor sprang from her body. The name Pegasus comes from the Greek word πηγαί (pegai), meaning 'springs' or 'waters'. Chrysaor's name means 'golden sword', in description of the blade he carried when he was born. Chrysaor played no further part in the story of Pegasus; he later became father of Geryon, the three-bodied monster whom Heracles slew.

Pegasus stretched his wings and flew away from the body of his mother, eventually arriving at Mount Helicon in Boeotia, home of the Muses. There, he struck the ground with his hoof and, to the delight of the Muses, from the rock gushed a spring of water which was named Hippocrene, 'horse's fountain'. The goddess Athene later came to see it.

Pegasus and Bellerophon

Pegasus is sometimes depicted as the steed of Perseus, but this is wrong. Pegasus was, in fact, ridden by another hero, Bellerophon, son of Glaucus. King Iobates of Lycia sent Bellerophon on a mission to kill the Chimaera, a fire-breathing monster that was devastating Lycia. According to Hesiod the Chimaera was the offspring of Typhon and Echidne, and had three heads, one like a lion, another like a goat and the third like a dragon. But Homer said in the *Iliad* that it had the front of a lion, the tail of a snake and a middle like a goat, the description that most other authors have followed.

Bellerophon found Pegasus drinking at the spring of Peirene in Corinth and tamed him with a golden bridle given by Athene. Ascending into the sky on the divine horse, Bellerophon swooped down on the Chimaera, killing it with arrows and a lance. After undertaking various other tasks for King Iobates, Bellerophon seems to have got over-inflated ideas, for he attempted to fly up on Pegasus to join the gods on Olympus. Before he got there he fell back to Earth; but Pegasus completed the trip and Zeus used him for a while to carry

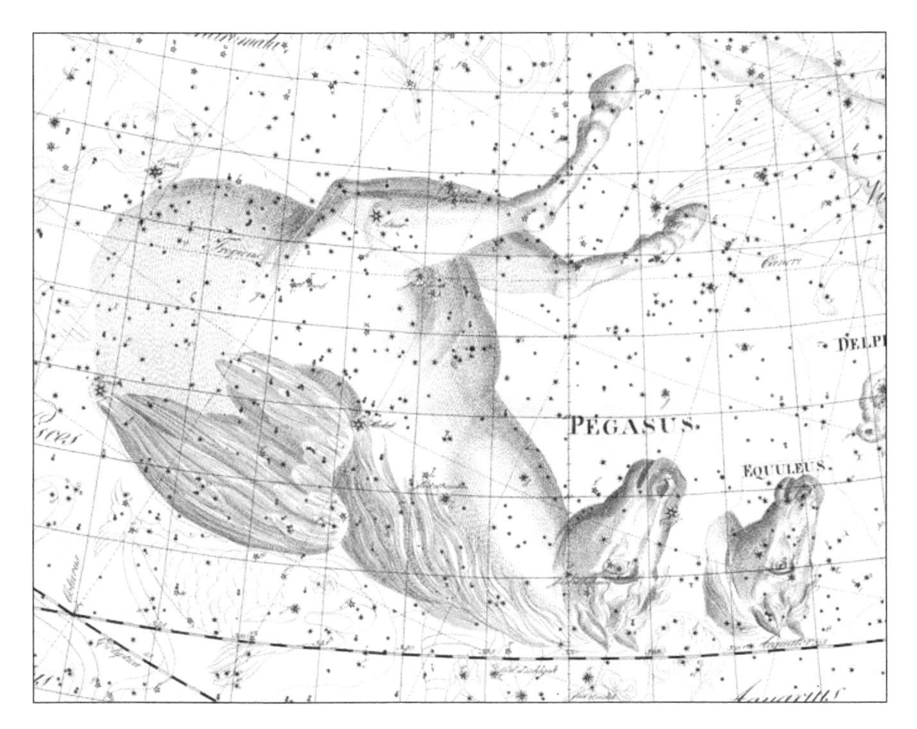

Only the front half of the flying horse is depicted in the sky, but enough to include his wings. His body is outlined by four stars that form the Square of Pegasus (although one of these is now assigned to Andromeda). In front of Pegasus is Equuleus, the foal (page 92), whose head alone is imagined among the stars. This illustration is from Chart X of the *Uranographia* of Johann Bode (1801). (Deutsches Museum)

his thunder and lightning, according to Hesiod. Zeus later put Pegasus among the constellations.

Eratosthenes doubted this story because, he said, the horse in the sky has no wings. It is true that Aratus did not mention wings on the celestial horse, but he identified the constellation as Pegasus, and Ptolemy in his *Almagest* definitely mentioned wings, so Eratosthenes was mistaken. Germanicus Caesar was in no doubt. Pegasus, he wrote, 'beats his swift wings in the topmost circle of the sky and rejoices in his stellification'. Eratosthenes repeated the claim of the 5th century BC playwright Euripides that this constellation represents Melanippe, daughter of Chiron the centaur (see Equuleus, page 93).

Pegasus in the sky

The Greeks from Aratus to Ptolemy knew the constellation simply as Ἵππος (Hippos), the horse; the horse was usually understood to be Pegasus, but only later did it become commonly known by that specific name. In the sky, only the top half of the horse is shown – even so, it is still the seventh-largest constellation. Its body is represented by the famous Square of Pegasus whose corners are marked by four stars. In Greek times, one star was considered common with

Andromeda, marking both the horse's navel and the top of Andromeda's head. When Johann Bayer came to assign Greek letters to the stars early in the 17th century he gave this a dual identity as both Alpha Andromedae and Delta Pegasi. Now it is allocated exclusively to Andromeda, as Alpha Andromedae; there is no longer a Delta Pegasi. Hence, strictly speaking, the Square of Pegasus now has only three corners.

The remaining three stars of the Square are: Alpha Pegasi, also known as Markab from the Arabic for 'shoulder'; Beta Pegasi, called Scheat from the Arabic meaning 'the shin'; and Gamma Pegasi, or Algenib, meaning 'the side' in Arabic. A star on the horse's muzzle, Epsilon Pegasi, is called Enif from the Arabic meaning 'nose'. Germanicus Caesar said it lies 'where the animal chews the bit, his mouth foaming'.

Perseus

Genitive: Persei
Abbreviation: Per
Size ranking: 24th
Origin: One of the 48 Greek constellations listed by Ptolemy in the *Almagest*
Greek name: Περσεύς (Perseus)

Perseus (Περσεύς in Greek) is one of the most famous Greek heroes. The characters of the story in which he is the action-man hero are represented by six constellations that occupy a substantial part of the sky. The constellation depicting Perseus himself lies in a prominent part of the Milky Way, which is perhaps why Aratus termed him 'dust-stained'.

In Greek myth, Perseus was the son of Danaë, daughter of King Acrisius of Argos. Acrisius had locked Danaë away in a heavily guarded dungeon when an oracle foretold that he would be killed by his grandson. But Zeus visited Danaë in the form of a shower of golden rain that fell through the skylight of the dungeon into her lap and impregnated her. When Acrisius found out, he locked Danaë and the infant Perseus into a wooden chest and cast them out to sea.

Inside the bobbing chest Danaë clutched her child and prayed to Zeus for deliverance from the sea. A few days later, the chest washed ashore on the island of Seriphos, its cargo still alive but starved and thirsty. A fisherman, Dictys, broke the chest open and found the mother and child. Dictys brought up Perseus as his own son.

The brother of Dictys was King Polydectes, who coveted Danaë as a wife. But Danaë was reluctant and Perseus, now grown to manhood, defended her from the king's advances. Instead, King Polydectes hatched a plan to get rid of Perseus. The king pretended he had turned his attentions to Hippodameia, daughter of King Oenomaus of Elis. King Polydectes asked his subjects, including Perseus, to provide horses for a wedding present. Perseus had no horse to give, nor money to buy one, so Polydectes sent him to bring the head of Medusa the Gorgon.

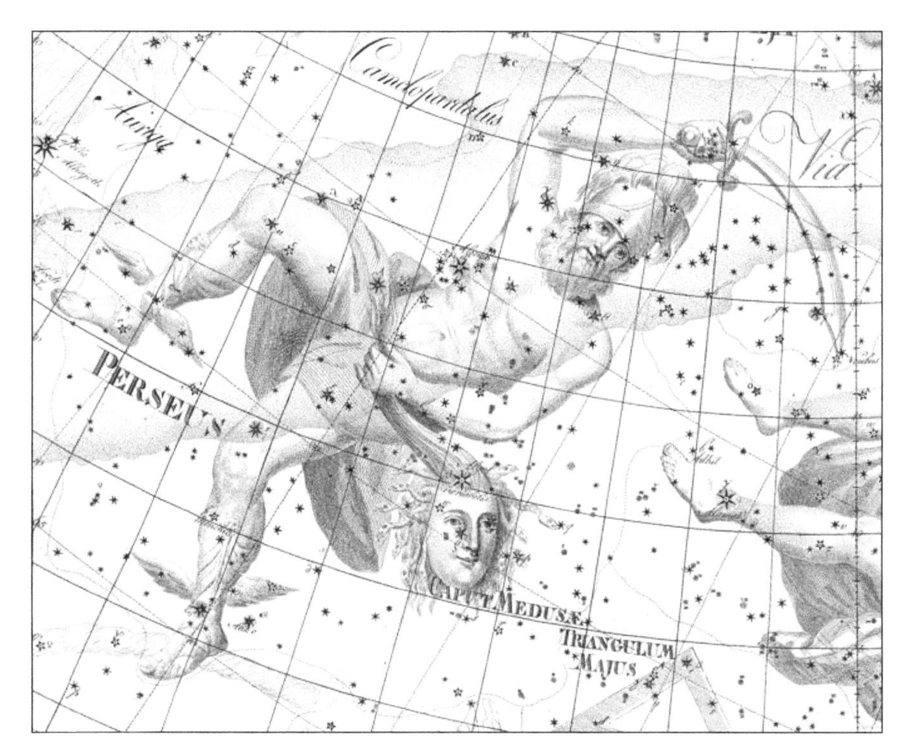

Perseus holding the severed head of Medusa the Gorgon, labelled Caput Medusae, as depicted on Chart IV of Johann Bode's *Uranographia* (1801). Perseus wears winged sandals and brandishes a curving scimitar. On the forehead of the Gorgon lies the star Algol, famous for its variations in light – Bode labelled it 'variabilis' on this chart. To the right lie the feet of Andromeda, whom he famously rescued. (Deutsches Museum)

Perseus and Medusa the Gorgon

The Gorgons were three hideously ugly sisters called Euryale, Stheno, and Medusa. They were the daughters of Phorcys, a god of the sea, and his sister Ceto. The Gorgons had faces covered with dragon scales, tusks like boars, hands of brass, and wings of gold. Their evil gaze turned to stone anyone who set eyes on them. Euryale and Stheno were immortal, but Medusa was mortal. She was distinguishable from the others because she had snakes for hair. In her youth Medusa had been famed for her beauty, particularly that of her hair, but she was condemned to a life of ugliness by Athene in whose temple she had been ravished by Poseidon.

A Gorgon's head would be a powerful weapon for a tyrannical king to enforce his rule, but King Polydectes probably thought that Perseus would die in his attempt to obtain it. However, the king had reckoned without Perseus's family connections among the gods. Athene gave him a bronze shield which he carried on his left arm, while in his right hand he wielded a sword of diamond made by Hephaestus. Hermes gave him winged sandals, and on his head he wore a helmet of darkness from Hades that made him invisible.

Perseus beheads Medusa

Under the guidance of Athene, Perseus flew to the slopes of Mount Atlas where the sisters of the Gorgons, called the Graeae, acted as lookouts. The Graeae were poorly qualified for the task, since they had only one eye between the three of them, which they passed to each other in turn. Perseus snatched the eye from them and threw it into Lake Tritonis.

He then followed a trail of statues of men and animals who had been turned to stone by the gaze of the Gorgons. Unseen in his helmet of invisibility, Perseus crept up on the Gorgons and waited until night when Medusa and her snakes were asleep. Looking only at her reflection in his brightly polished shield, Perseus swung his diamond sword and decapitated Medusa with one blow. As Medusa's head rolled to the ground, Perseus was startled to see the winged horse Pegasus and the armed warrior Chrysaor spring fully grown from her body, the legacy of her youthful affair with Poseidon. (Pegasus is commemorated in a constellation of its own.) Perseus rapidly collected up Medusa's head, put it in a pouch and flew away before the other Gorgons awoke.

Drops of blood fell from the head and turned into serpents as they struck the sands of Libya below. Strong winds blew Perseus across the sky like a raincloud, so he stopped to rest in the kingdom of Atlas. When Atlas refused him hospitality, Perseus took out the Gorgon's head and turned him into the range of mountains that now bear his name.

Perseus rescues Andromeda

The following morning Perseus resumed his flight with new vigour, coming to the land of King Cepheus whose daughter Andromeda was being sacrificed to a sea monster. Perseus's rescue of the girl, one of the most famous themes of mythology, is told in detail under the entry for Andromeda (page 37). Perseus returned triumphantly with Andromeda to the island of Seriphos, where he found his mother and Dictys sheltering in a temple from the tyranny of King Polydectes. Perseus stormed into the King's palace to a hostile reception. Reaching into his pouch, Perseus brought out the head of Medusa, turning Polydectes and his followers to stone. Perseus appointed his stepfather Dictys as king of Seriphos. Athene took the head of Medusa and set it in the middle of her shield.

Incidentally, the prophecy that had started all these adventures – namely, that Acrisius would be killed by his grandson – eventually came to pass during an athletics contest when a discus thrown by Perseus accidentally hit Acrisius, one of the spectators, and killed him. Perseus and Andromeda had many children, including Perses, whom they gave to Cepheus to bring up. From Perses, the kings of Persia were said to have been descended.

Algol the Demon and other stars in Perseus

In the sky, Perseus lies next to his beloved Andromeda. Nearby are her parents Cepheus and Cassiopeia, as well as the monster, Cetus, to which she was sacrificed. Pegasus the winged horse completes the tableau. Perseus himself is shown grasping the Gorgon's head by the hair with his left hand. In his right, he holds aloft his star-whetted sword, sometimes depicted as a curving scimitar.

Ptolemy described four stars as lying in the Gorgon's head; we know them as Beta, Omega, Rho, and Pi Persei. Several map-makers in the 17th, 18th, and 19th centuries made these and surrounding stars into a sub-constellation, Caput Medusae, the head of Medusa. Bode gave it a separate label on his *Uranographia* atlas (see illustration on page 141).

The star that Ptolemy called 'the bright one in the Gorgon head' is Beta Persei, named Algol from the Arabic *ra's al-ghul* meaning 'the demon's head'. Algol is the type of star known as an eclipsing binary, consisting of two close stars that orbit each other, in this case every 2.9 days. Algol varies in brightness as the two stars eclipse each other. Its variability was discovered in 1669 by the Italian astronomer Geminiano Montanari (1633–87).

It is sometimes speculated that the name Algol arose because the Arabs knew of its variability, but in fact the name has its origins in Greek mythology and its variability is simply coincidence. The Arabic astronomer al-Ṣūfī, who paid more attention to star brightnesses than had Ptolemy, made no mention of any variability in his *Book of the Fixed Stars* published in AD 964. The sketch of the constellation in his book places Algol next to Medusa's eye, not actually on it; many later cartographers, including Flamsteed and Bode, positioned the star even higher on the forehead. The idea of Algol being a 'winking eye' seems to be a modern myth.

The brightest star in the constellation, second-magnitude Alpha Persei, is officially named Mirfak, which comes from the Arabic for 'elbow'. A former alternative name was Algenib from the Arabic meaning 'the side', which is where Ptolemy described it as lying; however, this name is now officially assigned to Gamma Pegasi. The right hand of Perseus, in which he holds his sword, is marked by a feature that Ptolemy in the *Almagest* termed a 'nebulous mass' – in fact, a twin cluster of stars now known as the Double Cluster.

Phoenix

The phoenix

Genitive: Phoenicis
Abbreviation: Phe
Size ranking: 37th
Origin: The 12 southern constellations of Keyser and de Houtman

A constellation representing the mythical bird that supposedly was reborn from its own ashes. It is the largest of the 12 figures invented at the end of the 16th century by the Dutch navigators Pieter Dirkszoon Keyser and Frederick de Houtman. As with all the Keyser and de Houtman constellations it was first depicted on a globe by Petrus Plancius in 1598 and first appeared in print in 1603 on the *Uranometria* atlas of Johann Bayer. The constellation lies near the southern end of the river Eridanus and its brightest star, Alpha Phoenicis, is of magnitude 2.4.

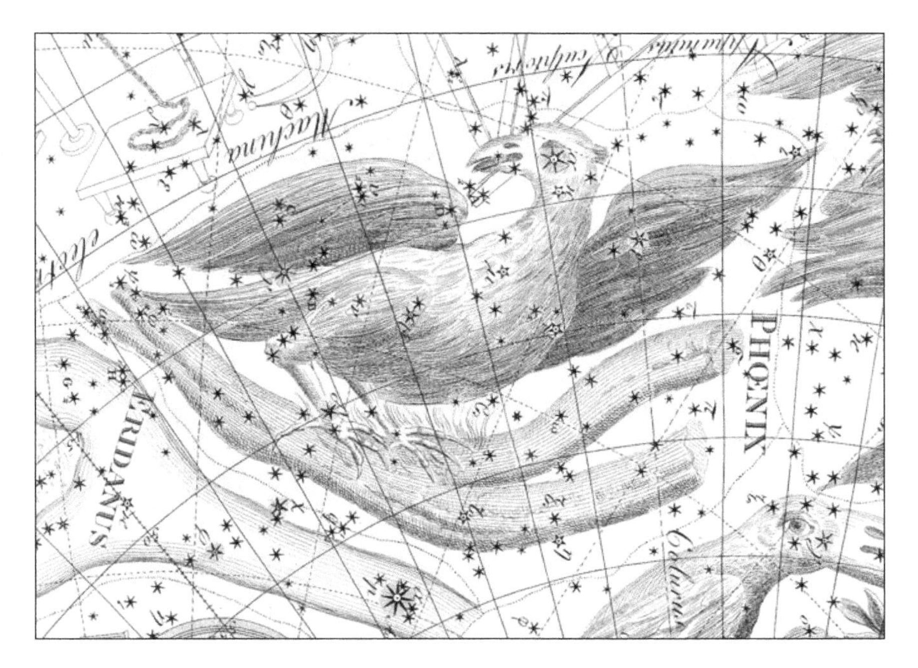

Phoenix, the multicoloured bird that ended its life on a funeral pyre, depicted on Chart XX of the *Uranographia* of Johann Bode (1801). (Deutsches Museum)

The phoenix was said to resemble a large eagle with scarlet, blue, purple, and gold plumage. Ovid in his *Metamorphoses* tells us that the phoenix lived for 500 years, eating the gum of incense and the sap of balsam. At the end of its allotted span the bird built itself a nest from cinnamon bark and incense among the topmost branches of a palm tree, ending its life on the fragrantly scented nest. A baby phoenix was born from its father's body. The nest was both the tomb of one phoenix and the cradle of the next.

When it was old enough to bear the weight, the young phoenix lifted the nest from the tree and carried it to the temple of Hyperion, the Titan who was the father of the Sun god. The death and rebirth of the phoenix has been seen as symbolizing the daily rising and setting of the Sun.

Why name a constellation after the phoenix when, bar Triangulum Australe, all the other 12 southern inventions of Keyser and de Houtman were based on real living creatures? Perhaps the answer lies with the exotic bird of paradise. When the first colourful specimens, all dead, arrived in Europe in the 16th century, speculation arose that they might be the mythical phoenix, or at least relatives thereof.

The Portuguese called them *passaros da sol*, birds of the Sun. The 16th-century French naturalist Pierre Belon believed in the possible existence of the phoenix and even gave it an entry in his 1555 book *L'histoire de la nature des oyseaux*. So there was good authority for the reality of such a bird at the end of the 16th century, when the constellation of the phoenix first entered the sky, even though neither it nor the bird of paradise had yet been seen alive by Europeans.

Pictor

The painter's easel

Genitive: Pictoris
Abbreviation: Pic
Size ranking: 59th
Origin: The 14 southern constellations of Nicolas Louis de Lacaille

One of the constellations representing technical and artistic apparatus which were introduced into the southern sky by the Frenchman Nicolas Louis de Lacaille after his observing expedition to the Cape of Good Hope in 1751–52. Pictor lies under the keel of the now-dismembered Greek constellation Argo Navis, the ship of the Argonauts, next to the bright star Canopus.

Lacaille's original title for the constellation, as given on his planisphere of 1756, was le Chevalet et la Palette, the easel and palette. In 1763 he Latinized this to Equuleus Pictorius (*sic*), while Johann Bode in his *Uranographia* of 1801 termed it Pluteum Pictoris, as illustrated below. In 1844 the English astronomer John Herschel proposed shortening the name to Pictor. This suggestion was adopted by Francis Baily in his *British Association Catalogue* of 1845 and it has been known as Pictor ever since.

Pictor, shown under the name Pluteum Pictoris on Chart XX of the *Uranographia* star atlas of Johann Bode (1801). Bode closely followed Lacaille's original depiction of this constellation, unlike in many other cases. The bright star at centre right is Canopus on the steering oar of the ship Argo. (Deutsches Museum)

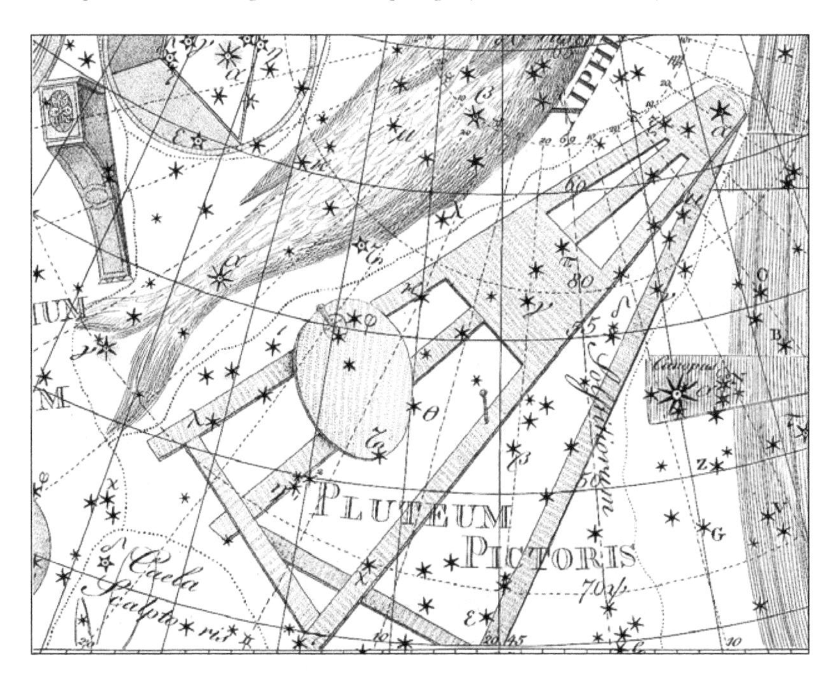

Pisces

The fishes

Genitive: Piscium
Abbreviation: Psc
Size ranking: 14th
Origin: One of the 48 Greek constellations listed by Ptolemy in the *Almagest*
Greek name: Ἰχθύες (Ichthyes)

The mythological events concerning this constellation are said to have taken place around the Euphrates river, a strong indication that the Greeks inherited this constellation from the Babylonians. The story follows an early episode in Greek mythology, in which the gods of Olympus had defeated the Titans and the Giants in a power struggle. Mother Earth, also known as Gaia, had another nasty surprise in store for the gods. She coupled with Tartarus, the lowest region of the Underworld where Zeus had imprisoned the Titans, and from this unlikely union came Typhon, the most awful monster the world had ever seen.

A cord joins the tails of Pisces, the two fish, as depicted in the *Atlas Coelestis* of John Flamsteed (1729). The horizontal dashed line passing through the southerly fish is the celestial equator, and the diagonal dashed line is the Sun's annual path, the ecliptic. The point where they cross is known as the vernal equinox. (University of Michigan Library)

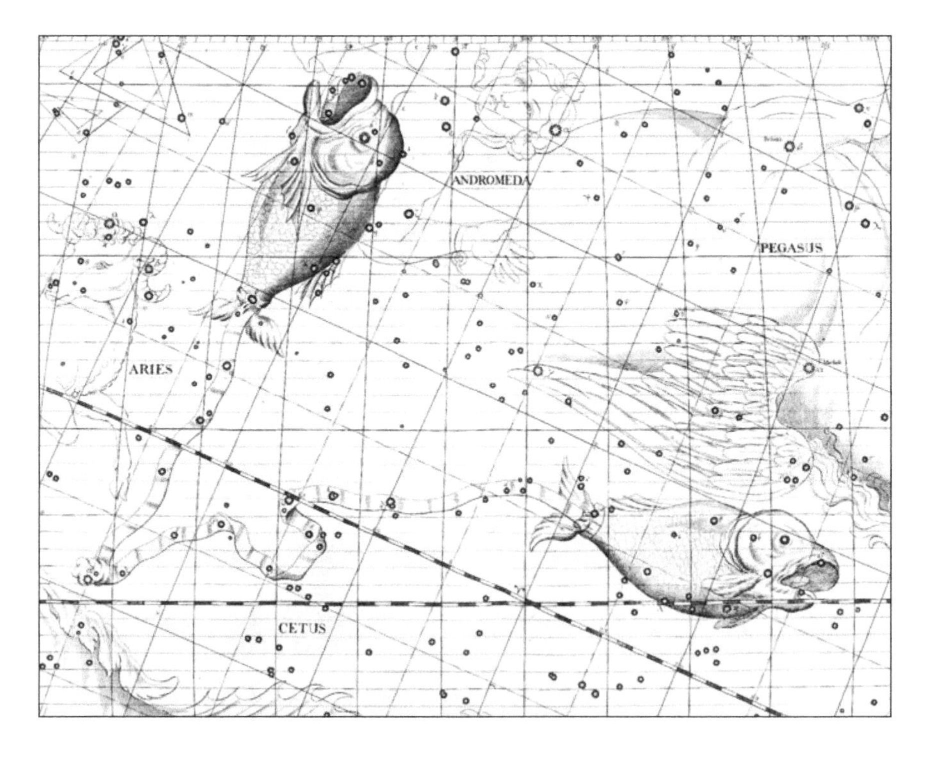

According to Hesiod in the *Theogony*, Typhon had a hundred dragon's heads from which black tongues flicked out. Fire blazed from the eyes in each of these heads, and from them came a cacophony of sound: sometimes ethereal voices which only the gods could understand, while at other times Typhon bellowed like a bull, roared like a lion, yelped like puppies, or hissed like a nest of snakes.

Gaia sent this fearsome monster to attack the gods. Pan saw him coming and alerted the others with a shout. Pan himself jumped into the river and changed his form into a goat-fish, represented by the constellation Capricornus, also inherited from the Babylonians.

The goddess Aphrodite and her son Eros took cover among the reeds on the banks of the Euphrates, but when the wind rustled the undergrowth Aphrodite became fearful. Holding Eros in her lap she called for help to the water nymphs and leapt into the river. In one version of the story, two fish swam up and carried Aphrodite and Eros to safety on their backs, although in another version the two refugees were themselves changed into fish. The mythologists said that because of this story the Syrians would not eat fish, regarding them as gods or the protectors of gods.

An alternative story, given by Hyginus in the *Fabulae*, is that an egg fell into the Euphrates and was rolled to the shore by two fish. Doves sat on the egg and from it hatched Aphrodite who, in gratitude, put the fish in the sky. Eratosthenes had yet another explanation: he wrote that the two fish represented by Pisces were offspring of the much larger fish that is represented by the constellation Piscis Austrinus. When the goddess Derceto fell into a lake near Bambyce in northern Syria, she was rescued by the large fish; she placed this fish and its two youngsters in the sky as Piscis Austrinus and Pisces, respectively.

The Greek name for the constellation was Ἰχθύες (Ichthyes); Pisces is the Latin equivalent. In the sky, the two fish of Pisces are represented swimming at right angles to each other, one northwards and the other westwards, their tails joined by a cord or ribbon. The Greeks offered no good explanation for this cord, but according to the historian Paul Kunitzsch the Babylonians visualized a pair of fish joined by a cord in this area, so evidently the Greeks borrowed this idea although the significance of the cord was lost.

Stars of Pisces, and the vernal equinox

Pisces is a disappointingly faint constellation, its brightest stars being of only fourth magnitude. Alpha Piscium is called Alrescha, from the Arabic name meaning 'the cord'. Ptolemy described this star as lying where the cords joining the two fish are knotted together.

Pisces is notable because it contains the point at which the Sun crosses the celestial equator into the northern hemisphere each year on March 20. This point, called the vernal equinox, lay in Aries in Greek times and as a result is also known as the first point of Aries. It has since moved into Pisces because of a slow wobble of the Earth on its axis called precession, which causes celestial coordinates to drift slowly over time. The continuing effect of precession will carry the vernal equinox into Aquarius around the year 2597.

Piscis Austrinus

The southern fish

Genitive: Piscis Austrini
Abbreviation: PsA
Size ranking: 60th
Origin: One of the 48 Greek constellations listed by Ptolemy in the *Almagest*
Greek name: Ἰχθύς Νότιος (Ichthys Notios)

Eratosthenes called this the Great Fish and said that it was the parent of the two smaller fish of the zodiacal constellation Pisces. Like Pisces, its mythology has a Middle Eastern setting that reveals its Babylonian origin. According to the brief account of Eratosthenes, the Syrian fertility goddess Derceto (the Greek name for Atargatis) is supposed to have fallen into a lake at Bambyce near the river Euphrates in northern Syria, and was saved by a large fish. Hyginus said, in repetition of his note on Pisces, that as a result of this the Syrians do not eat fish but they worship the images of fish as gods. All the accounts of this constellation's mythology are disappointingly sketchy.

Bambyce later became known to the Greeks as Hieropolis (meaning 'sacred city'), now called Manbij. Other classical sources tell us that temples of Atargatis

Piscis Austrinus, called Piscis Notius on Chart XVI of Johann Bode's *Uranographia* (1801), is visualized as lying on its back and drinking water flowing from the urn of Aquarius. In its mouth is the bright star Fomalhaut. (Deutsches Museum)

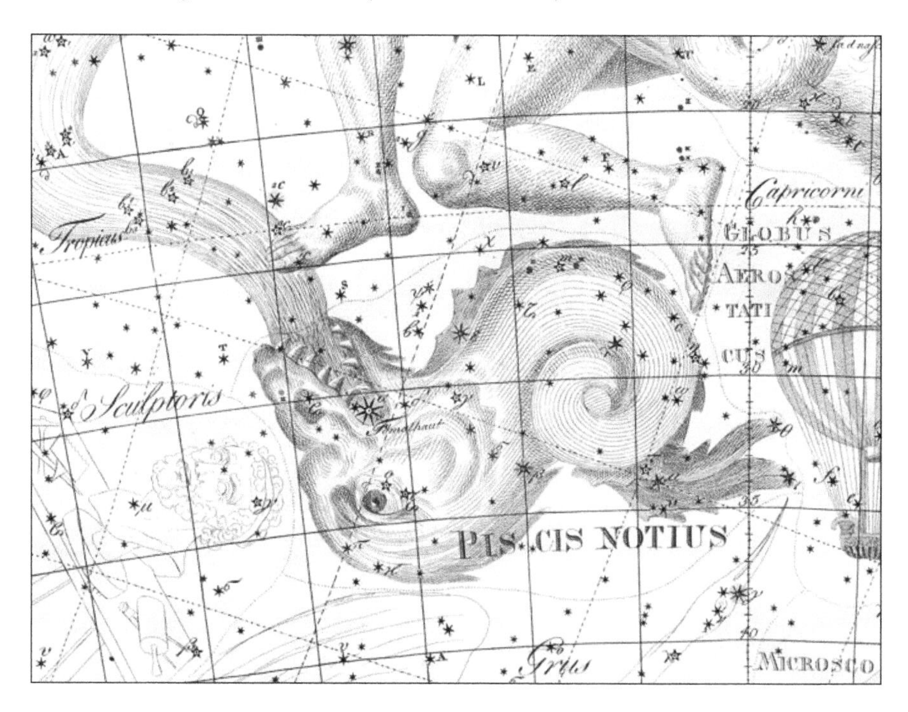

contained ponds of sacred fish. The goddess was said to punish those who ate fish by making them ill, but her priests safely ate fish in a theophagic ritual.

According to the Greek writer Diodorus Siculus (1st century BC), Derceto deliberately threw herself into a lake at Ascalon in Palestine as a suicide bid in shame for a love affair with a young Syrian, Caystrus, by whom she bore a daughter, Semiramis. Derceto killed her lover and abandoned her child, who was brought up by doves and later became queen of Babylon. In the lake, Derceto was turned into a mermaid, half woman, half fish.

Ptolemy listed this constellation in the *Almagest* as Ἰχθύς Νότιος (Ichthys Notios), while a common Latin alternative was Piscis Notius, used by Bayer, Hevelius, and Bode. John Flamsteed, though, preferred the title Piscis Austrinus in his star catalogue (1725) and atlas (1729) and his choice eventually prevailed.

Piscis Austrinus is more noticeable in the sky than the zodiacal Pisces because it contains the first-magnitude star Fomalhaut. This name comes from the Arabic *fam al-ḥūt* meaning 'fish's mouth', which is where Ptolemy described it as lying. In the sky the fish is shown drinking the water flowing from the jar of Aquarius, a strange thing for a fish to do. The flow of water ends at Fomalhaut, which Ptolemy regarded as being common to Aquarius and Piscis Austrinus. The name Fomalhaut is sometimes mis-spelled 'Formalhaut'.

In the *Almagest*, Ptolemy listed six additional stars in this area that did not form part of Piscis Austrinus; these are now assigned to the modern figure of Microscopium (page 124). In addition, when the 12 new southern constellations of Keyser and de Houtman were invented at the end of the 16th century, the star that Ptolemy placed at the tip of the fish's tail was appropriated for use as the head of the new constellation Grus, the crane. Piscis Austrinus was the final constellation in Ptolemy's star catalogue.

Puppis

The stern

Genitive: Puppis
Abbreviation: Pup
Size ranking: 20th
Origin: Part of the original Greek constellation Argo Navis

The largest of the three sections into which the ancient constellation of Argo Navis, the ship of the Argonauts, was divided by Nicolas Louis de Lacaille in his catalogue of the southern stars published in 1756. In that catalogue he gave it the French name Pouppe du Navire. His final catalogue, *Coelum Australe Stelliferum*, appeared in 1763 containing the same subdivisions but with Latin rather than French names. See page 187 for the full story of Argo.

Puppis represents the stern, or poop, of the ship; the other sections were Carina, the keel or hull, and Vela, the sails. Puppis has no stars labelled Alpha or Beta (and nor does Vela). This is a result of Lacaille's relettering of the stars in Argo. Instead of lettering each of the three sections as separate constellations, he treated them as though they were still part of a single figure. Alpha, Beta, and Epsilon were allocated to stars in the subdivision of Carina, while Gamma and Delta went to stars in Vela. The brightest star in Puppis is in fact second-magnitude Zeta Puppis, called Naos from the Greek word for ship, ναῦς.

Pyxis
The compass

Genitive: Pyxidis
Abbreviation: Pyx
Size ranking: 65th
Origin: The 14 southern constellations of Nicolas Louis de Lacaille

A small southern constellation invented by the Frenchman Nicolas Louis de Lacaille during his survey of the southern skies in 1751–52. Pyxis represents a magnetic compass as used by seamen. It is located near the stern of the ship Argo in the same area as the ship's mast. Lacaille first published it on his chart of 1756 under the French name la Boussole, but Latinized it to Pixis Nautica (*sic*) on the second edition of his chart in 1763. The name was subsequently

Pyxis hovers over the mast of Argo on Chart XVIII of the *Uranographia* of Johann Bode (1801). Curling around it is one of Bode's inventions, the now-obsolete Lochium Funis, the log and line (page 196). (Deutsches Museum)

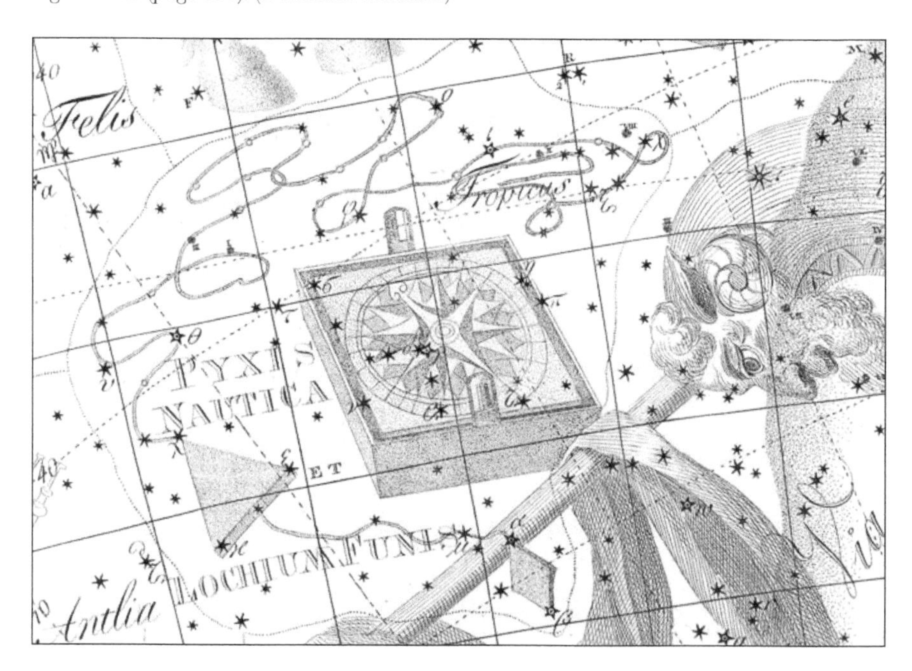

shortened, with amended spelling, to just Pyxis. The brightest stars in Pyxis are of only fourth magnitude and there are no legends associated with it – indeed, the magnetic compass was completely unknown to the ancient Greeks.

It is sometimes asserted that Pyxis is one of the parts into which Lacaille divided the former Argo Navis, but that is not the case. Lacaille split Argo into three: Carina (Corps), Puppis (Pouppe), and Vela (Voilure). Pyxis is an additional constellation invented by Lacaille, and he showed it separately from Argo on his map and in his catalogue.

The four stars that Lacaille labelled Alpha, Beta, Gamma, and Delta Pyxidis had been catalogued by Ptolemy as lying in the mast of Argo. Johann Bayer did not assign Greek letters to them in his *Uranometria* atlas of 1603 so Lacaille evidently felt free to appropriate them for his new constellation.

In 1844 the English astronomer John Herschel proposed returning these stars to Argo by replacing Pyxis with a fourth subdivision of the ship which he called Malus, the mast, in deference to Ptolemy's original description. His countryman Francis Baily included Malus in his *British Association Catalogue* of 1845, but otherwise Malus was not widely adopted.

In this same area of sky the German astronomer Johann Bode introduced another constellation, Lochium Funis, the Log and Line (see page 196), shown coiling around Pyxis on Bode's atlas, opposite, but now obsolete.

Reticulum

The net

Genitive: Reticuli
Abbreviation: Ret
Size ranking: 82nd
Origin: The 14 southern constellations of Nicolas Louis de Lacaille

A small southern constellation, introduced by the French astronomer Nicolas Louis de Lacaille to commemorate the reticle in the eyepiece of his small telescope with which he measured star positions from the Cape of Good Hope in 1751–52. It consisted of a diamond shape formed by silk threads inserted into the eyepiece which helped him judge the position of stars as they passed through the field of view.

In the notes to his southern star catalogue published by the French Royal Academy of Sciences in 1756 Lacaille described it as 'The little instrument used to construct this catalogue. It is constructed by the intersection of four lines drawn from each corner of a square to the middle of the two opposite sides.' He originally gave it the French name le Réticule but Latinized this to Reticulus (*sic*) on the second edition of his chart in 1763; on that second chart he also labelled its stars with Greek letters. The name became Reticulum in Benjamin Gould's *Uranometria Argentina* catalogue of 1879. The constellation's brightest star, Alpha Reticuli, is of third magnitude, but is not named.

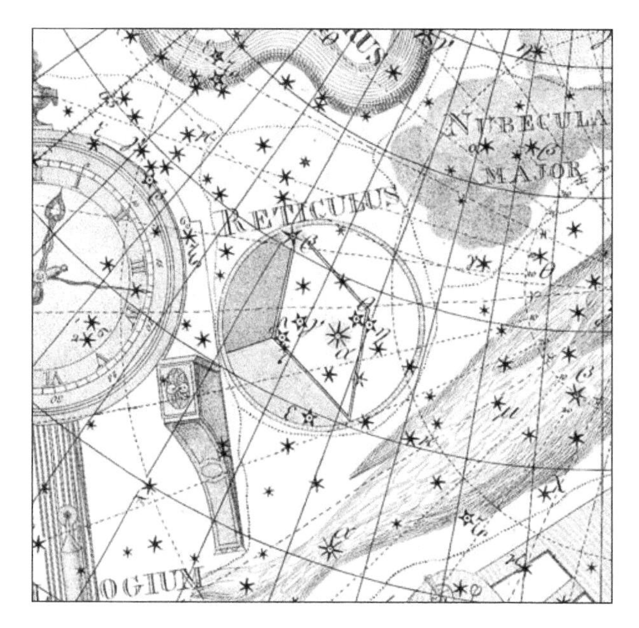

Reticulum, shown on Chart XX of Johann Bode's *Uranographia* under the name Reticulus, which is what Lacaille, its inventor, had called it. To the upper right is Nubecula Major, the Large Magellanic Cloud. (Deutsches Museum)

Lacaille's Reticulum replaced a previous constellation in this same area called Rhombus introduced in 1621 on a globe by the German astronomer Isaac Habrecht (1589–1633). Rhombus later appeared in print in Habrecht's *Planiglobium Coeleste, et Terrestre* published in 1628. Habrecht's Rhombus, though, was considerably larger than Lacaille's Reticulum and extended farther south. It consisted of the present-day Alpha and Beta Reticuli plus Gamma and Nu Hydri, forming a quadrilateral in the space between Dorado and Hydrus.

Sagitta

The arrow

Genitive: Sagittae
Abbreviation: Sge
Size ranking: 86th
Origin: One of the 48 Greek constellations listed by Ptolemy in the *Almagest*
Greek name: 'Οϊστός (Oistos)

Sagitta is the third-smallest constellation in the sky, and has no stars brighter than fourth magnitude, but it was well-known to the Greeks and was among the 48 constellations listed by Ptolemy in the *Almagest*. Aratus described it as 'alone, without a bow' since there is no sign of the archer who might have shot it. Aratus and Ptolemy knew it by the Greek name 'Οϊστός (Oistos), but puzzlingly Eratosthenes called it Τόξον (Toxon) which means 'bow' rather than 'arrow'. An alternative Latin name for the constellation widely used prior to the 18th century was Telum, meaning a dart or spear.

There are at least three different stories to account for the arrow in the sky. Eratosthenes said it was the projectile with which Apollo killed the Cyclopes because they made the thunderbolts of Zeus that struck down Apollo's son, Asclepius. According to this story, Asclepius was a great healer with the power to raise the dead, but Zeus killed Asclepius when Hades, god of the Underworld, complained that he was losing business. Asclepius is commemorated in the constellation Ophiuchus.

Hyginus, on the other hand, said that Sagitta was one of the arrows with which Heracles killed the eagle that ate the liver of Prometheus. It was Prometheus who moulded men out of clay in the likeness of the gods, and gave them fire that he had stolen from Zeus. Prometheus carried the fire triumphantly in a vegetable stalk like a runner bearing the Olympic torch. Zeus cruelly punished him for this theft by chaining him to Mount Caucasus, where a long-winged eagle ate his liver during the day. But at night the liver grew again for the eagle to resume his feast in the morning. Heracles freed Prometheus from this eternal torture by shooting the eagle with an arrow.

Germanicus Caesar identified Sagitta as the arrow of Eros which kindled in Zeus his passion for the shepherd boy Ganymede, who is commemorated by the constellation Aquarius. Now, according to Germanicus, the arrow is guarded in the sky by the eagle of Zeus – and Sagitta does indeed lie next to the constellation of the eagle, Aquila.

Alpha Sagittae, magnitude 4.4, has the strange-sounding name Sham, which comes from the Arabic name for the constellation, *al-sahm*, meaning 'the arrow'. The constellation's brightest star is actually Gamma Sagittae, magnitude 3.5, which Ptolemy described as lying on the arrow-head; however, the atlases of Flamsteed and Bode extended the shaft beyond Gamma to Eta Sagittae, as in the illustration below, a star that Ptolemy did not list.

Sagitta flying under the forefeet of Vulpecula, from Chart VIII of the *Uranographia* of Johann Bode (1801). The constellation's brightest star, Gamma Sagittae, lies where the right forepaw of Vulpecula touches the arrow's shaft. (Deutsches Museum)

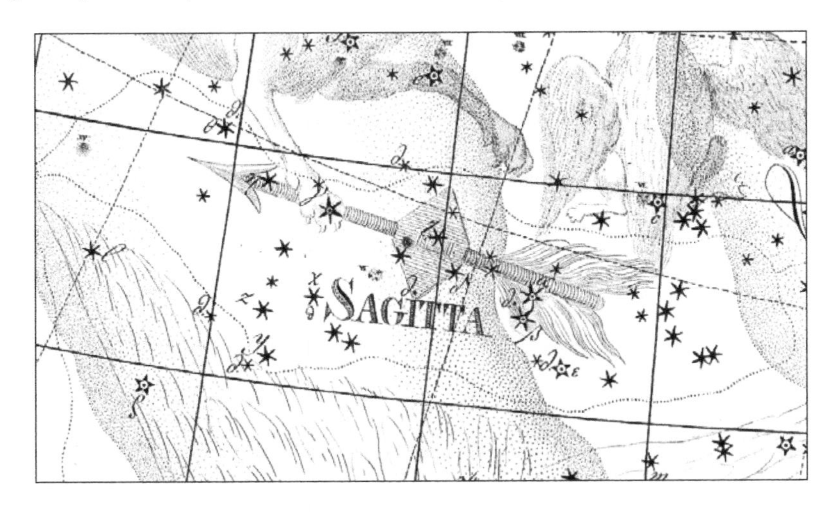

Sagittarius
The archer

Genitive: Sagittarii
Abbreviation: Sgr
Size ranking: 15th
Origin: One of the 48 Greek constellations listed by Ptolemy in the *Almagest*
Greek name: Τοξότης (Toxotes)

Sagittarius is depicted in the sky as a centaur, with the body and four legs of a horse but the upper torso of a man. He is shown wearing a cloak and drawing a bow, aimed in the direction of the neighbouring scorpion, Scorpius. Aratus spoke of the Archer, Τοξότης (Toxotes), and his Bow, Τόξον (Toxon), as though they were separate constellations. Most likely this is because the stars of the bow and arrow are the most distinctive part of the figure. They form the asterism that we now know as the Teapot (see below).

Sagittarius is a constellation of Sumerian origin that represented PA.BIL.SAG, a god of war and hunting who they depicted as a centaur-like archer with wings. The Sumerian figure was subsequently adopted by the Greeks, although without the wings. As a result there are no particular myths associated with this constellation and the Greek mythographers were confused as to its identity.

Some doubted that this was a centaur at all, among them Eratosthenes who gave as one of his reasons the fact that centaurs did not use bows. Instead, Eratosthenes described Sagittarius as a two-legged creature with the tail of a satyr. He said that this figure was Crotus, son of Eupheme, the nurse to the Muses, who were nine daughters of Zeus. The Roman mythographer Hyginus in his *Fabulae* added the information that his father was Pan, agreeing with Eratosthenes that the archer was a satyr rather than a centaur.

Crotus was said to have invented archery and often went hunting on horseback. He lived on Mount Helicon among the Muses, who enjoyed his company. They sang for him, and he applauded them loudly. The Muses requested that Zeus place him among the stars, where he is seen demonstrating the art of archery. In the sky he was given the hind legs of a horse because he was such a keen horseman.

Aratus and Ptolemy, though, both spoke of the archer as four-legged, which is how he is usually depicted. Ptolemy described him with a flowing cloak, known as the ephaptis, attached at his shoulders. By his forefeet is a circle of stars that Hyginus said was a wreath 'thrown off as by one at play'. This circlet of stars is the constellation Corona Australis (see page 80). Sagittarius was sometimes misidentified as Chiron, a wise and scholarly centaur, but Chiron is in fact represented by the other celestial centaur, the constellation Centaurus.

Stars of Sagittarius

Alpha Sagittarii is called Rukbat, from the Arabic *rukbat al-rami*, 'knee of the archer'. Beta Sagittarii is called Arkab, from the Arabic name meaning 'the

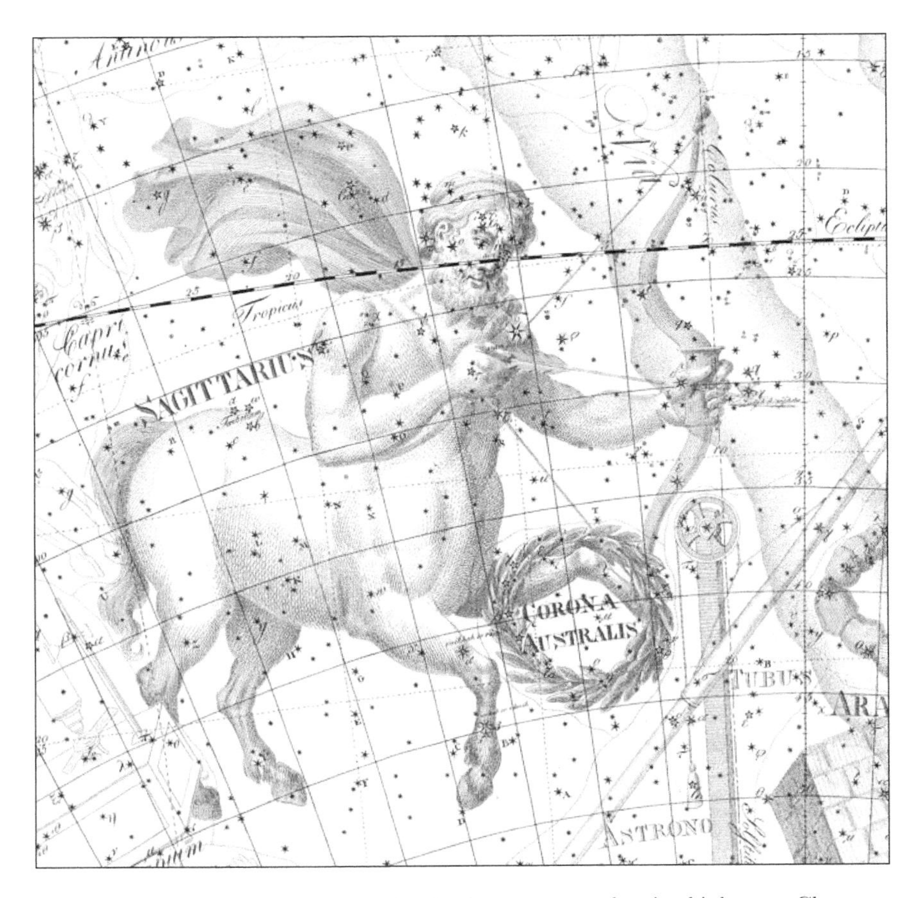

Sagittarius, the centaur-like archer with a flowing cape, seen drawing his bow on Chart XV of the *Uranographia* of Johann Bode (1801). Between his forelegs lies Corona Australis, the southern crown, shown here as a wreath. (Deutsches Museum)

archer's Achilles tendon'. Gamma Sagittarii is Alnasl, from the Arabic meaning 'the point', referring to the tip of the archer's arrow.

Delta, Epsilon, and Lambda Sagittarii are respectively called Kaus Media, Kaus Australis, and Kaus Borealis. The word Kaus comes from the Arabic *al-qaus*, 'the bow', while the suffixes are Latin words signifying the middle, southern, and northern parts of the bow. Zeta Sagittarii is Ascella, a Latin word meaning 'armpit'. All these names closely follow the descriptions of the stars' positions given by Ptolemy in his *Almagest*.

Last, but not least, is Sigma Sagittarii, called Nunki. This name was applied relatively recently by navigators, but it was borrowed from a list of Babylonian star names. The Babylonian name NUN-KI was given to a group of stars representing their sacred city of Eridu on the Euphrates. The name is now applied to Sigma Sagittarii alone, and is reputedly the oldest star name in use.

Ptolemy in the *Almagest* inexplicably classified the stars that we know as Alpha and Beta Sagittarii as second magnitude, when they are in fact fourth. Bayer, who lived too far north to see these stars for himself, accepted Ptolemy's assessment and labelled them Alpha and Beta. In fact, Alpha Sagittarii is only the 15th brightest star in the constellation, over seven times fainter than the brightest star, Epsilon, which is magnitude 1.8.

Tea, with milk

Among present-day astronomers, the shape outlined by the eight main stars of Sagittarius (Gamma, Delta, Epsilon, Lambda, Phi, Sigma, Tau, and Zeta) is popularly known as the Teapot. Its handle consists of Phi, Sigma, Tau, and Zeta, the top of the lid is marked by Lambda, while Delta, Epsilon, and Gamma are the triangular spout. This same group of stars, with the addition of Mu Sagittarii, was originally visualized as the archer's bow and arrow. A subset of these stars – Lambda, Phi, Sigma, Tau, and Zeta – form a ladle shape called the Milk Dipper, fittingly placed in a rich area of the Milky Way.

Sagittarius contains dense Milky Way star fields that lie towards the centre of our Galaxy. The exact centre of the Galaxy is believed to be marked by a radio-emitting source that astronomers call Sagittarius A, near the border with Ophiuchus and close to the point of the archer's arrow. There are many notable objects in Sagittarius, including the Lagoon Nebula and the Trifid Nebula, two much-photographed clouds of gas lit up by stars inside them.

Scorpius
The scorpion

Genitive: Scorpii
Abbreviation: Sco
Size ranking: 33rd
Origin: One of the 48 Greek constellations listed by Ptolemy in the *Almagest*
Greek name: Σκορπίος (Skorpios)

'There is a certain place where the scorpion with his tail and curving claws sprawls across two signs of the zodiac', wrote Ovid in his *Metamorphoses*. He was referring to the ancient Greek version of Scorpius, which was much larger than the constellation we know today. The Greek scorpion was divided into two halves: one half, called Σκορπίος (Skorpios), contained its body and sting, while the front half comprised the claws. The Greeks called this front half Χηλαί (Chelae), which means 'claws'. In the first century BC the Romans made the claws into a separate constellation, Libra, the balance.

In mythology, this is the scorpion that stung Orion the hunter to death, although accounts differ as to the exact circumstances. Eratosthenes offers two versions. Under his description of Scorpius he says that Orion tried to ravish Artemis, the hunting goddess, and that she sent the scorpion to sting him, an account that is supported by Aratus. But in his entry on Orion, Eratosthenes says that the Earth sent the scorpion to sting Orion after he had boasted that

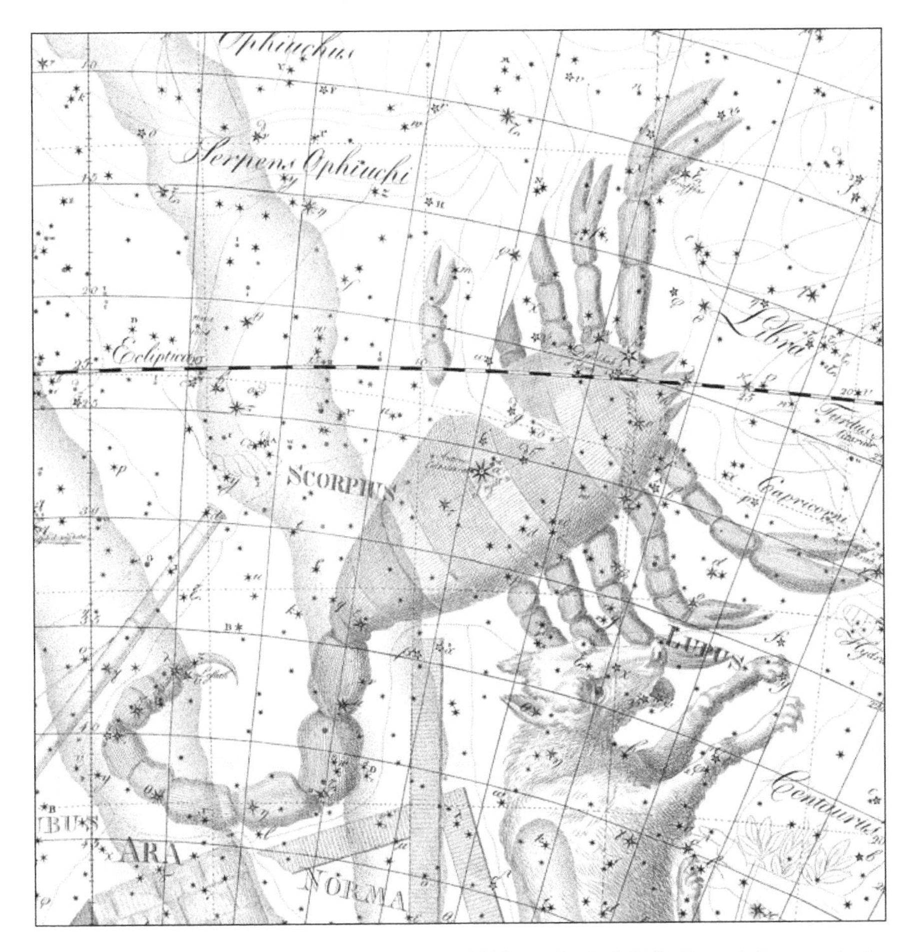

Scorpius from Chart XV of the *Uranographia* of Johann Bode (1801). Part of the scorpion's body is overlapped from above by the left leg and foot of Ophiuchus. In the middle of the scorpion's body lies the bright, reddish star Antares to which Bode also gives the alternative name Calbalakrab, from the Arabic meaning 'scorpion's heart'. (Deutsches Museum)

he could kill any wild beast. Hyginus also gives both stories. Aratus says that the death of Orion happened on the island of Chios, but Eratosthenes and Hyginus place it in Crete.

In either case, the moral is that Orion suffers retribution for his hubris. This seems to be one of the oldest of Greek myths and its origin may lie in the sky itself, since the two constellations are placed opposite each other so that Orion sets as his conqueror the scorpion rises. But the constellation is much older than the Greeks, for the Sumerians knew it as GIR-TAB, meaning the scorpion, over 5,000 years ago.

Scorpius clearly resembles the shape of a scorpion, particularly the curving line of stars that form its tail with its sting raised to strike. Old star maps such as the one above show the lower left leg and foot of Ophiuchus, to the north, awkwardly overlapping the scorpion's body. Incidentally, Scorpius is the modern astronomical name for the constellation; Scorpio is the old name, now used only by astrologers.

Stars of Scorpius

The brightest star in Scorpius is brilliant Antares, a name that comes from the Greek word Ἀντάρης meaning 'like Mars', on account of its strong reddish-orange colour, similar to that of the planet Mars. The name is often translated as 'rival of Mars', but the star name expert Paul Kunitzsch prefers the translation 'like Mars'. Antares is a remarkable supergiant star, several hundred times the diameter of our Sun, lying about 550 light years away.

Beta Scorpii is officially named Acrab, from the Arabic for 'scorpion'; an obsolete alternative was Graffias, Latin for 'claws'. Delta Scorpii is called Dschubba, a strange-sounding name that is a corruption of the Arabic word meaning 'forehead', in reference to its position in the middle of the scorpion's head. At the end of the scorpion's tail lies Lambda Scorpii, called Shaula from the Arabic meaning 'the sting' which is what Ptolemy said that it marked.

Ptolemy in the *Almagest* listed three stars as lying outside the constellation (i.e. they were so-called unformed stars). The first of these he described as 'the nebulous star to the rear of the sting'. This is most likely the large and bright open cluster we know as M7, which as a result is sometimes called Ptolemy's Cluster. M7 is the most southerly of the objects catalogued by the French astronomer Charles Messier, at declination –34.8°.

Sculptor

The sculptor

Genitive: Sculptoris
Abbreviation: Scl
Size ranking: 36th
Origin: The 14 southern constellations of Nicolas Louis de Lacaille

A faint constellation south of Cetus and Aquarius, invented by the French astronomer Nicolas Louis de Lacaille during his mapping of the southern skies in 1751–52. His original name for it, given on his planisphere of 1756, was l'Atelier du Sculpteur, the sculptor's studio. It consisted of a carved head on a tripod table, with the artist's mallet and two chisels on a block of marble next to it. On Lacaille's 1763 planisphere the title was Latinized to Apparatus Sculptoris. Johann Bode in 1801 dispensed with the block of marble and moved the sculptor's tools to the top of the table along with the carved bust, as seen on the facing page. In place of the marble block he created the constellation Machina Electrica (page 197), but that figure never achieved wide currency.

In 1844 the English astronomer John Herschel proposed shortening the constellation's name to Sculptor. This suggestion was adopted by Francis Baily

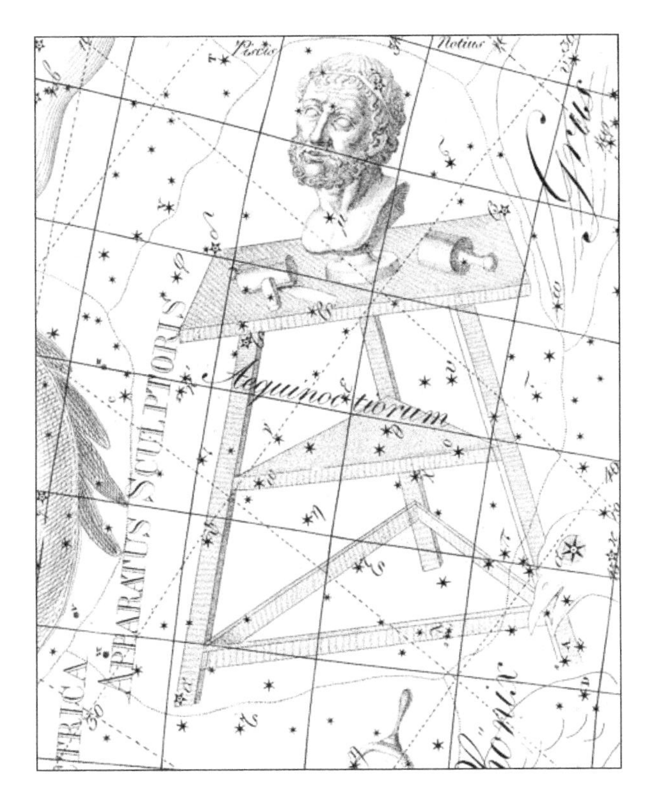

Sculptor, shown under the name Apparatus Sculptoris, on Chart XVII of the *Uranographia* of Johann Bode (1801). As originally depicted by its inventor, Lacaille, the sculptor's studio included a block of marble in front of the table. Bode replaced the marble block with a new constellation, Machina Electrica (off the bottom of this illustration), but it did not survive. (Deutsches Museum)

in his *British Association Catalogue* of 1845, and the constellation has been known simply as Sculptor ever since. Sculptor is the largest of Lacaille's 14 inventions, as defined by the modern constellation boundaries, but its stars are only of fourth magnitude and fainter; none of them is named.

Scutum

The shield

Genitive: Scuti
Abbreviation: Sct
Size ranking: 84th
Origin: The seven constellations of Johannes Hevelius

The fifth-smallest constellation in the sky, introduced in 1684 by the Polish astronomer Johannes Hevelius under the title Scutum Sobiescianum, Sobieski's Shield. He named it in honour of King John III Sobieski of Poland who helped Hevelius rebuild his observatory after a disastrous fire in 1679.

Hevelius's description and chart of the constellation first appeared in August 1684 in *Acta Eruditorum*, a leading scientific journal of the day. Hevelius quoted Edmond Halley's invention six years earlier of Robur Carolinum, honouring

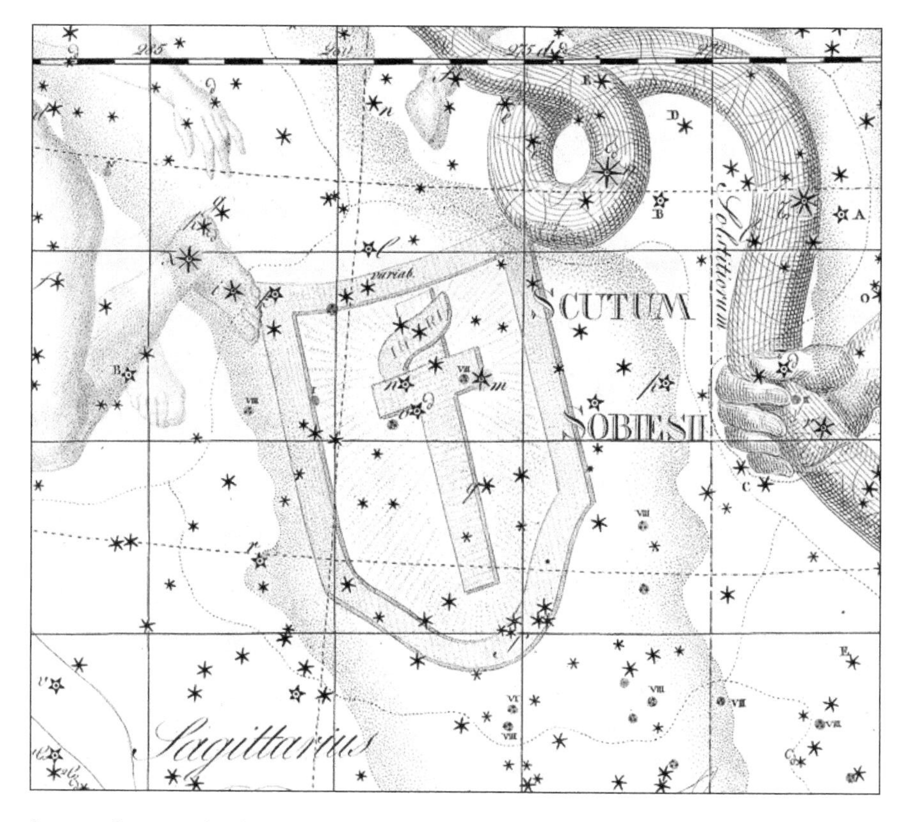

Scutum shown under the name Scutum Sobiesii on Chart IX of the *Uranographia* of Johann Bode (1801). It is squeezed between the feet of the now-obsolete Antinous (left), the tail of Serpens (above and right), and Sagittarius (below). (Deutsches Museum)

King Charles II of England, as the precedent. Robur Carolinum did not survive but Scutum did, and is in fact the only constellation introduced for political reasons that is still in use.

Even so, Scutum nearly didn't make it. John Flamsteed ignored it in his catalogue and atlas, listing its stars under Aquila, although he accepted six other Hevelius inventions. Johann Bode reinstated it to the sky in his *Uranometria* of 1801 under the name Scutum Sobiesii (see illustration above). It was taken out again by the English astronomer Francis Baily who omitted it from his influential *British Association Catalogue* of 1845. This tug-of-war was resolved by the American astronomer Benjamin Gould, who included it as plain Scutum in his *Uranometria Argentina* catalogue of 1879 and allocated Greek letters to its stars for the first time, thereby cementing its permanence.

Scutum lies in a bright area of the Milky Way between Aquila to the north and Sagittarius to the south and is distinctive despite its small size. Its brightest stars are of only fourth magnitude, and none is named, but the constellation contains a celebrated cluster of stars popularly known as the Wild Duck cluster because its fan-shape resembles a flight of ducks.

Serpens

The serpent

Genitive: Serpentis
Abbreviation: Ser
Size ranking: 23rd
Origin: One of the 48 Greek constellations listed by Ptolemy in the *Almagest*
Greek name: Ὄφις (Ophis)

This constellation is unique, for it is divided into two parts – Serpens Caput, the head, and Serpens Cauda, the tail. Nevertheless, astronomers regard it as a single constellation. Serpens represents a huge snake held by the constellation Ophiuchus. Its usual Greek name was Ὄφις (Ophis), but in the *Almagest* Ptolemy gave it as Ὄφις Ὀφιούχου, i.e. the serpent of the serpent-holder, presumably to prevent confusion with the other celestial serpents Draco and Hydra.

In his left hand Ophiuchus grasps the top half of the snake, while his right hand holds the tail. Aratus and Manilius agreed that Serpens was coiled around the body of Ophiuchus, but most star atlases show the snake simply passing between his legs or behind his body (for an illustration of the full tableau, see Ophiuchus, page 131).

In mythology Ophiuchus was identified as the healer Asclepius, a son of Apollo, although why he appears to be wrestling with a serpent in the sky is not fully explained. His connection with snakes is attributed to the story that he once killed a snake which was miraculously restored to life by a herb placed on it by another snake. Asclepius subsequently used the same technique to revive dead people. Snakes are the symbol of rebirth because they shed their skins every year.

Ptolemy in the *Almagest* listed the stars of Ophiuchus and Serpens as separate constellations, but mythologists such as Eratosthenes and Hyginus dealt with them as a unit. One of the few celestial map makers to show Serpens separately was Johann Bayer, who devoted a page of his atlas to each of the Ptolemaic constellations. Other major celestial chart makers such as Hevelius, Flamsteed, and Bode followed the mythologists in treating Ophiuchus and the snake as a composite figure, with the snake partly obscured by the body of Ophiuchus.

When Eugène Delporte came to define the official constellation boundaries in the late 1920s he was faced with the problem of how to separate the intertwined figures of Ophiuchus and Serpens. His solution was to chop Serpens into the two halves that we see today, the head on one side of Ophiuchus and the tail on the other. The head is by far the larger half, covering an area over twice that of the tail. From the western border to the east the official boundaries of Serpens span nearly 57° of sky, just over half the length of Hydra, the water snake, which is the largest constellation of all.

The constellation's brightest star, third-magnitude Alpha Serpentis, is called Unukalhai from the Arabic meaning 'the serpent's neck', where it is located. The tip of the serpent's tail is marked by Theta Serpentis, called Alya, an Arabic word that actually refers to a sheep's tail. The most celebrated object in Serpens is a star cluster called M16, embedded in a gas cloud called the Eagle Nebula, which takes its name from its supposed resemblance to a large bird of prey. The Eagle Nebula was the subject of a famous photograph by the Hubble Space Telescope showing pillars of gas and forming stars.

Sextans
The sextant

Genitive: Sextantis
Abbreviation: Sex
Size ranking: 47th
Origin: The seven constellations of Johannes Hevelius

A faint constellation south of Leo, introduced by the Polish astronomer Johannes Hevelius in his catalogue and atlas of 1687 under the name Sextans Uraniae. It commemorated the instrument he used for measuring star positions, which was destroyed along with other instruments in a fire at his observatory in 1679. Hevelius had continued to make naked-eye sightings with his sextant throughout his life, even though he used telescopes for observing the Moon and planets; it was perhaps to demonstrate the keenness of his eyes that he formed Sextans out of such faint stars, as he also did with another of his inventions, Lynx.

John Flamsteed shortened the constellation's name to Sextans in his *Catalogus Britannicus* of 1725 and on his accompanying *Atlas Coelestis* (see the illustration). Francis Baily also called it simply Sextans in his *British Association Catalogue* of

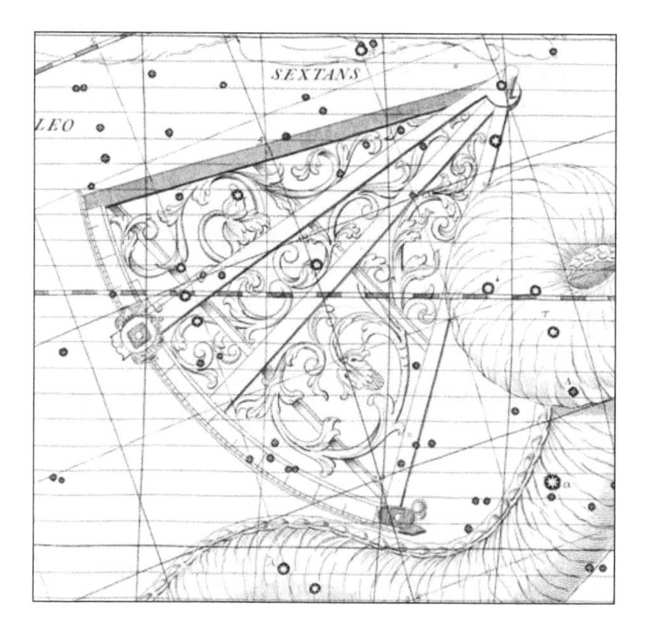

Sextans above the coils of Hydra, illustrated in the *Atlas Coelestis* of John Flamsteed (1729). Leo the lion lies to the north. (University of Michigan Library)

1845, but did not allocate Greek letters to any of its stars because none of them is brighter than 5th magnitude. That job was done later by the American astronomer Benjamin Gould in his *Uranometria Argentina* catalogue of 1879. The brightest star in the constellation, Alpha Sextantis, is magnitude 4.5. None of the stars is named.

Taurus
The bull

Genitive: Tauri
Abbreviation: Tau
Size ranking: 17th
Origin: One of the 48 Greek constellations listed by Ptolemy in the *Almagest*
Greek name: Ταῦρος (Tauros)

Taurus (Ταῦρος in Greek) is a distinctive constellation, with star-tipped horns and a head defined by a V-shaped group of stars. Two Greek bull-myths are associated with Taurus. Usually it was said to represent Zeus in the disguise he adopted for another of his extramarital affairs, this time as the bull that carried away Europa, daughter of King Agenor of Phoenicia.

Europa liked to play on the beach with the other girls of Tyre. Zeus instructed his son Hermes to drive the king's cattle from their pastures on the mountain slopes towards the shore where the girls were playing. Adopting the shape of a bull, Zeus surreptitiously mingled with the lowing herd, awaiting his chance to abduct Europa. There was no mistaking who was the most handsome bull. His hide was white as fresh snow and his horns shone like polished metal.

Europa was entranced by this beautiful yet placid creature. She adorned his horns with flowers and stroked his flanks, admiring the muscles on his neck and the folds of skin on his flanks. The bull kissed her hands, while inwardly Zeus could hardly contain himself in anticipation of the final conquest. The bull lay on the golden sands and Europa ventured to sit on his back. At first, she feared nothing when the bull rose and began to paddle in the surf. But she became alarmed when it began to swim strongly out to sea. Europa looked around in dismay at the receding shoreline and clung tightly to the bull's horns as waves washed over the bull's back. Craftily, Zeus the bull dipped more deeply into the water to make her hold him more tightly still.

By now, Europa had realized that this was no ordinary bull. Eventually, the bull waded ashore at Crete, where Zeus revealed his true identity and seduced her. He gave her presents that included a dog that later became the constellation Canis Major. The offspring of Zeus and Europa included Minos, king of Crete, who established the famous palace at Knossos where bull games were held.

An alternative story says that Taurus may represent Io, another illicit love of Zeus, whom the god turned into a heifer to disguise her from his wife Hera. But Hera was suspicious and set the hundred-eyed watchman Argus to guard the heifer. Hera, furious at this, sent a gadfly to chase the heifer, who threw herself into the sea and swam away.

In the sky, only the front half of the bull is shown. This can be explained mythologically by assuming that the hind quarters are submerged. In reality, there is no space in the sky to show the complete bull, for it is too big. The constellations Cetus and Aries lie where the bull's hind quarters would otherwise be. Taurus shares with Pegasus this uncomfortable fate of having been sliced in half in the sky. Adding to the awkwardness, Taurus moves across the sky backwards, as though retreating from Orion.

Aratus described the bull as 'crouching', and star maps have traditionally depicted Taurus with its front legs folded, perhaps lowering itself to entice Europa onto its back. Manilius described the bull as lame and drew a mawkish moral from it: 'The sky teaches us to undergo loss with fortitude, since even constellations are fashioned with limbs deformed', he wrote.

The Hyades – the face of the bull

The face of Taurus is marked by the V-shaped group of stars called the Hyades (Ὑάδες in Greek). Ovid in his *Fasti* asserts that the name comes from the old Greek word *hyein*, meaning 'to rain', so that Hyades means 'rainy ones', because their rising at certain times of year was said to be a sign of rain. In mythology the Hyades were the daughters of Atlas and Aethra the Oceanid. Their eldest brother was Hyas, a bold hunter who one day was killed by a lioness. His sisters wept inconsolably – Hyginus says they died of grief – and for this they were placed in the sky. Hence it seems equally likely that their name comes from their brother Hyas. In another story, the Hyades were nymphs who nursed the infant Dionysus in their cave on Mount Nysa, feeding him on milk and honey. The Romans had a different name: they called the Hyades *suculae* meaning 'piglets'.

The mythographers were massively confused about the names and even the number of the Hyades. They are variously described as being five or seven in number. Ptolemy listed five Hyades in his star catalogue. Hyginus alone gives four different lists of their names, none of which agrees completely with the list of five given originally by Hesiod, namely Phaesyle, Coronis, Cleia, Phaeo, and Eudore. Astronomers have avoided the problem by not naming any of the stars of the Hyades.

The Hyades is a genuine cluster of stars lying about 150 light years away. Binoculars and small telescopes show many more members of the Hyades than are visible to the naked eye. In all, the cluster is estimated to contain several hundred stars.

The Pleiades – seven celestial sisters

Even more famous than the Hyades is another star cluster in Taurus: the Pleiades (Πλειάδες in Greek), commonly known as the Seven Sisters. To a casual glance, the Pleiades cluster appears as a fuzzy patch like a swarm of flies over the back of the bull. According to Hyginus, some ancient astronomers called them the bull's tail. So distinctive are the Pleiades that the ancient Greeks regarded them as a separate mini-constellation and used them as a calendar

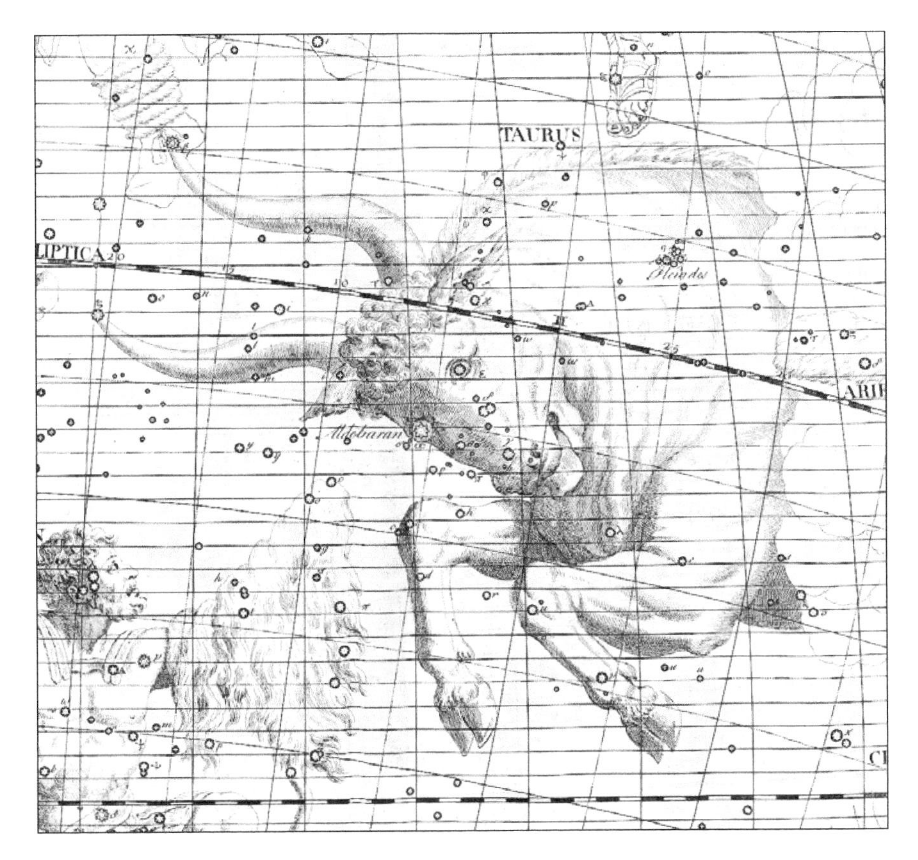

Taurus charges with head down towards Orion, as depicted in the *Atlas Coelestis* of John Flamsteed (1729). Only the front part of the bull is shown in the sky. The bull's eye is marked by the reddish star Aldebaran, while on his back is the Pleiades star cluster. One horn ends at the right foot of Auriga. (University of Michigan Library)

marker. Hesiod, in his agricultural poem *Works and Days*, instructs farmers to begin harvesting when the Pleiades rise at dawn, which in Greek times would have been in May, and to plough when they set at dawn, which would have been in November. Ptolemy did not list individual members of the Pleiades in his *Almagest*, giving only an indication of the cluster's size.

In mythology the Pleiades were the seven daughters of Atlas and the oceanid Pleione, after whom they are named. One popular derivation is that the name comes from the Greek word *plein*, meaning 'to sail' – so Pleione means 'sailing queen' and the Pleiades are the 'sailing ones', because in Greek times they were visible all night during the summer sailing season. When the Pleiades vanished from the night sky, it was considered prudent to remain ashore. 'Gales of all winds rage when the Pleiades, pursued by violent Orion, plunge into the clouded sea', wrote Hesiod.

Alternatively, and possibly more likely, the name may come from the old Greek word *pleos*, 'full', which in the plural meant 'many', a suitable reference to the cluster. According to other authorities, the name comes from the Greek word *peleiades*, meaning 'flock of doves'.

Unlike their half-sisters the Hyades, the names of all seven Pleiades are assigned to stars in the cluster: Alcyone, Asterope, Celaeno, Electra, Maia, Merope, and Taygeta. Two more stars are named after their parents, Atlas and Pleione. Alcyone is the brightest star in the cluster. According to mythology, Alcyone and Celaeno were both seduced by Poseidon. Maia, the eldest and most beautiful of the sisters, was seduced by Zeus and gave birth to Hermes; she later became foster-mother to Arcas, son of Zeus and Callisto. Zeus also seduced two others of the Pleiades: Electra, who gave birth to Dardanus, the founder of Troy; and Taygeta, who gave birth to Lacedaemon, founder of Sparta. Asterope was ravished by Ares and became mother of Oenomaus, king of Pisa, near Olympia, who features in the legend of Auriga. Hence six Pleiades became paramours of the gods. Only Merope married a mortal, Sisyphus, a notorious trickster who was subsequently condemned to roll a stone eternally up a hill.

Although the Pleiades are popularly termed the Seven Sisters, only six stars are easily visible to the naked eye, and a considerable mythology has grown up to account for the 'missing' Pleiad. Eratosthenes says that Merope was the faint Pleiad because she was the only one who married a mortal. Hyginus and Ovid also recount this story, giving her shame as the reason for her faintness, but both add another candidate: Electra, who could not bear to see the fall of Troy, which had been founded by her son Dardanus. Hyginus says that, moved by grief, she left the Pleiades altogether, but Ovid says that she merely covered her eyes with her hand. Astronomers, however, have not followed either legend in their naming of the stars, for the faintest named Pleiad is actually Asterope.

Binoculars show dozens of stars in the Pleiades, and in all the cluster contains a hundred or so members. The Pleiades lie about 440 light years away, nearly three times the distance of the Hyades. They are relatively youthful by stellar standards, the youngest being no more than a few million years old.

A famous myth links the Pleiades with Orion. As Hyginus tells it, Pleione and her daughters were one day walking through Boeotia when Orion tried to ravish her. Pleione and the girls escaped, but Orion pursued them for seven years. Zeus immortalized the chase by placing the Pleiades in the heavens where Orion follows them endlessly.

The eye, the horns – and a nebula named the Crab

The bull's glinting red eye is marked by the brightest star in Taurus, Aldebaran, a name that comes from the Arabic *al-dabarān* meaning 'the follower'; according to the 10th-century Arabic astronomer al-Ṣūfī, this name arose because it follows the Pleiades across the sky. Surprisingly for such a prominent star, Greek astronomers had no name for it (although Ptolemy called it Torch in his *Tetrabiblos*, a book about astrology). Aldebaran appears to be a member of the Hyades but in fact is a foreground object at less than half the distance, and so is superimposed on the cluster by chance. It is a red giant star about 40 times the diameter of the Sun. Aldebaran marks the right eye of the bull; the left eye is represented by Epsilon Tauri, with Gamma Tauri on the nose.

At the tip of the bull's left horn is Beta Tauri, or Elnath, a name that comes from the Arabic meaning 'the butting one'. Ptolemy described this star as being common with the right foot of Auriga, the charioteer, but since the introduction of rigorously defined constellation boundaries in 1930 it is now the exclusive property of Taurus. Hence the bull has kept the tip of his horn, but the charioteer has lost his right foot.

Near the tip of the bull's right horn, which is marked by Zeta Tauri, lies the remarkable Crab Nebula, the result of one of the most celebrated events in the history of astronomy – a stellar explosion, seen from Earth in AD 1054, that was bright enough to be visible in daylight for three weeks. We now know this this event was a supernova, the violent death of a massive star, and the Crab Nebula is the shattered remnant of the star that blew up, now visible only through telescopes. The Irish astronomer Lord Rosse (1800–67) gave the nebula its name in 1848 because he thought its shape resembled a crab.

Telescopium

The telescope

Genitive: Telescopii
Abbreviation: Tel
Size ranking: 57th
Origin: The 14 southern constellations of Nicolas Louis de Lacaille

One of the faint and obscure constellations of the southern sky introduced by the Frenchman Nicolas Louis de Lacaille after his sky-mapping trip to the Cape of Good Hope in 1751–52. It represents the type of long, unwieldy refractor

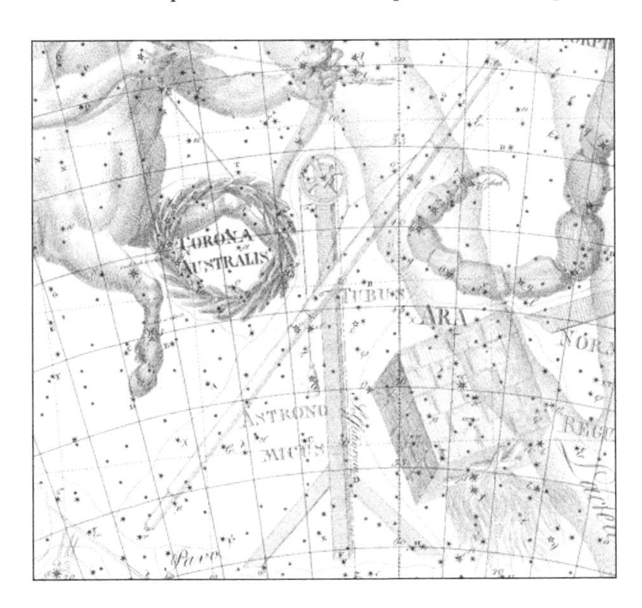

Telescopium, shown under the name Tubus Astronomicus on Chart XV of the *Uranographia* by Johann Bode (1801). It was envisaged as a long-tubed refractor operated by ropes and pulleys. The shape of the modern constellation is quite different, the upper part of the tube and mounting having been completely cut off. (Deutsches Museum)

suspended from a pole known as an aerial telescope, as used by J. D. Cassini at Paris Observatory. The reason for the great length was to reduce chromatic aberration (false colour) produced by the crude lenses of that time.

Lacaille originally depicted Telescopium as extending northwards between Sagittarius and Scorpius, as shown on the accompanying illustration by Johann Bode, but modern astronomers have cut off the top of the telescope's tube and mounting so that it is now restricted to a rectangular area of sky south of Sagittarius and Corona Australis.

As a result, Lacaille's Beta Telescopii, positioned in the pulley at the top of the mast, is now Eta Sagittarii; Gamma Telescopii, in the upper part of the refractor's tube, is G Scorpii; and Lacaille's Theta Telescopii, which marked the telescope's objective lens, is humble 45 Ophiuchi. Its brightest star, Alpha Telescopii, is magnitude 3.5.

Triangulum

The triangle

Genitive: Trianguli
Abbreviation: Tri
Size ranking: 78th
Origin: One of the 48 Greek constellations listed by Ptolemy in the *Almagest*
Greek name: Τρίγωνον (Trigonon)

Since any three points make up the corners of a triangle it is unsurprising, if somewhat unimaginative, to find a triangle among the constellations. Aratus and Eratosthenes knew this constellation as Δελτωτόν (Deltoton), because its

Triangulum from the *Atlas Coelestis* of John Flamsteed (1729). South of it lies a smaller triangle, Triangulum Minus, now obsolete (see page 207). (University of Michigan Library)

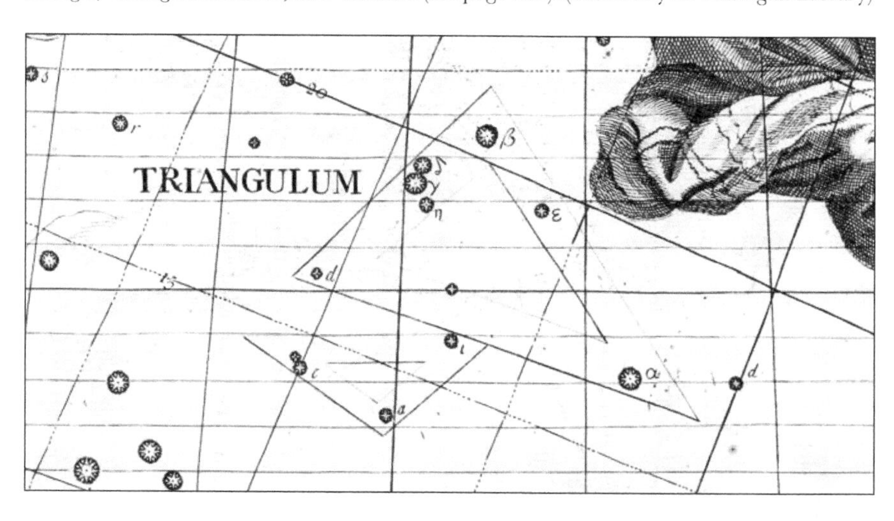

shape resembled a capital delta (Δ), while Ptolemy in the *Almagest* listed it as Τρίγωνον (Trigonon), triangle.

Aratus described it as an isosceles triangle, having two equal sides and a shorter third side. Eratosthenes said that it represented the Nile river delta. According to Hyginus, some people also saw it as the island of Sicily, which was originally known as Trinacria on account of its three promontories. In mythology, Trinacria was the home of Ceres, goddess of agriculture.

Alpha Trianguli, magnitude 3.4, is named Mothallah, which comes from the Arabic name for the constellation, *al-muthallath*, meaning 'the triangle'. The brightest star in Triangulum is actually Beta Trianguli, magnitude 3.0, but this has no name. Triangulum contains M33, a galaxy in our Local Group sometimes popularly termed the Pinwheel Galaxy, visible with binoculars.

A lesser triangle

A smaller triangle, Triangulum Minus, was introduced in 1687 by the Polish astronomer Johannes Hevelius who formed it from three stars next to Triangulum. He renamed the existing Triangulum as Triangulum Majus to distinguish it from the smaller sibling. Triangulum Minus was shown on some charts, such as the one reproduced here, but has since fallen into disuse (see also page 207).

Triangulum Australe

The southern triangle

Genitive: Trianguli Australis
Abbreviation: TrA
Size ranking: 83rd
Origin: The 12 southern constellations of Keyser and de Houtman

One of the 12 constellations introduced at the end of the 16th century by the Dutch navigators Pieter Dirkszoon Keyser and Frederick de Houtman, and the smallest of them according to modern boundaries. A southern triangle had previously been shown in a completely different position, south of Argo Navis, on a globe of 1589 by the Dutchman Petrus Plancius, along with a southern cross, but they were not the constellations we know today. The modern Triangulum Australe was first depicted in 1598 on a globe by Petrus Plancius and first appeared in print in 1603 on the *Uranometria* atlas of Johann Bayer.

The three main stars of Triangulum Australe are brighter than those of their northern counterpart, although the constellation is smaller. Navigators have named its brightest star Atria, a contraction of its scientific name Alpha Trianguli Australis.

On his 1756 planisphere of the southern stars the French astronomer Nicolas Louis de Lacaille referred to it as 'le Triangle Austral ou le Niveau' ('niveau' meaning level) and he even showed it with an attached plumb bob, indicating that he regarded it as representing a surveyor's level. 'Niveau' was later Latinized

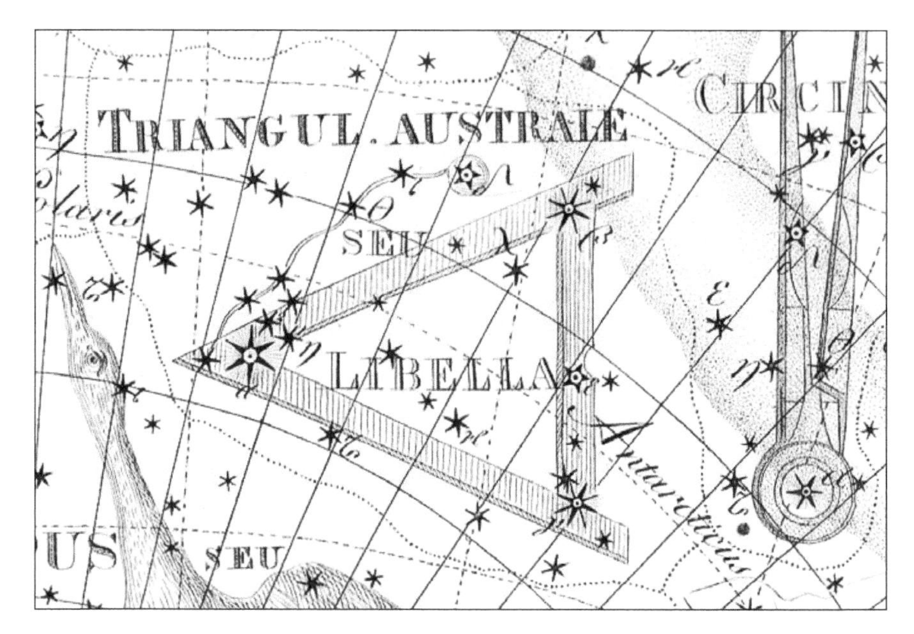

Triangulum Australe, with the alternative name Libella, the level, on Chart XX of Johann Bode's *Uranographia* (1801). Bode followed Lacaille in showing a plumb bob attached to the triangle, thereby representing it as a surveyor's level. Along with the compasses (Circinus, to the right) and a set square and ruler (Norma, out of picture at the top) it formed a toolkit of surveying instruments in this part of the sky. (Deutsches Museum)

to 'libella', as on Bode's atlas shown here. Through some misreading, the historian R. H. Allen transferred the appellation 'level' to the nearby constellation Norma and termed that constellation the Level and Square (instead of the Rule and Square), thereby confusing generations of astronomers.

Tucana

The toucan

Genitive: Tucanae
Abbreviation: Tuc
Size ranking: 48th
Origin: The 12 southern constellations of Keyser and de Houtman

Tucana is one of the 12 southern constellations devised by the Dutch navigators Pieter Dirkszoon Keyser and Frederick de Houtman at the end of the 16th century. It represents the South American bird with a huge bill.

The Dutchman Petrus Plancius gave it the name Toucan when he first depicted it on a globe in 1598, and Johann Bayer followed suit on his atlas of 1603. But de Houtman, in his catalogue of 1603, called it Den Indiaenschen

Exster, op Indies Lang ghenaemt ('the Indian magpie, named Lang in the Indies', the word 'lang' referring to the bird's long beak). De Houtman was apparently describing not a toucan but the hornbill, a similarly endowed bird that is native to the East Indies and Malaysia. This suggests that the original inventor of Tucana was in fact Keyser, who had visited South America before his voyage to the East Indies and could have seen the bird there. In some depictions which used de Houtman's catalogue as a source, such as Willem Janszoon Blaeu's globe of 1603, the bird was shown as a hornbill rather than a toucan, complete with casque above its bill, but the original identification as a toucan won out.

Tucana's brightest star, Alpha Tucanae, marking the tip of the bird's beak, is of only third magnitude, but the constellation is distinguished by two features of particular interest: the globular star cluster 47 Tucanae, rated the second-best such object in the entire sky, so bright that it was labelled in the same way as a star; and also the Small Magellanic Cloud, the smaller and fainter of the two companion galaxies of our Milky Way. These features were originally part of Hydrus but were transferred to Tucana by the French astronomer Nicolas Louis de Lacaille when he reorganized this part of the southern heavens in the 1750s.

Incidentally, 47 Tucanae is not a Flamsteed number; it comes from its listing in Johann Bode's catalogue called *Allgemeine Beschreibung und Nachweisung der Gestirne*, published in 1801 to accompany his *Uranographia* star atlas. It was recorded as a star by Keyser and de Houtman, but its nebulous nature was first noted by Lacaille a century and a half later. None of the stars of Tucana is named, and there are no legends associated with it.

Tucana, holding in its beak a branch with a berry, as seen on Chart XX of Johann Bode's *Uranographia* star atlas (1801). Beneath its tail lies Nubecula Minor, the Small Magellanic Cloud, which is within the borders of the modern constellation. (Deutsches Museum)

Ursa Major
The great bear

Genitive: Ursae Majoris
Abbreviation: UMa
Size ranking: 3rd
Origin: One of the 48 Greek constellations listed by Ptolemy in the *Almagest*
Greek name: Ἄρκτος Μεγάλη (Arktos Megale)

Undoubtedly the most familiar star pattern in the entire sky is the seven stars that make up the shape popularly termed the Plough or Big Dipper, part of the third-largest constellation in the sky, Ursa Major, the great bear. The seven stars form the rump and tail of the bear, while the rest of the animal is comprised of fainter stars. Its Greek name in the *Almagest* was Ἄρκτος Μεγάλη (Arktos Megale); Ursa Major is the Latin equivalent.

Aratus called the constellation Ἑλίκη (Helike), meaning 'twister', apparently from its circling of the pole, and said that the ancient Greeks steered their ships by reference to it. In the *Odyssey*, for example, we read that Odysseus kept the great bear to his left as he sailed eastwards. The Phoenicians, on the other hand, used the little bear (Ursa Minor), which Aratus termed Κυνόσουρα (Kynosoura, or Cynosura in Latin transliteration). Aratus tells us that the bears were also called wagons or wains, and in one place he referred to the figure of Ursa Major as the 'wagon-bear' to underline its dual identity.

Homer in the *Odyssey* referred to 'the Great Bear that men call the Wain, that circles opposite Orion, and never bathes in the sea', the last phrase being a reference to its circumpolar (non-setting) nature. The adjacent constellation Boötes was imagined as either the herdsman of the bear or the wagon driver. Germanicus Caesar seems to have been the first to mention a third, now-common identity – he said that the bears were also called ploughs because, as he wrote, 'the shape of a plough is the closest to the real shape formed by their stars'.

According to Hyginus the Romans referred to the great bear as Septentrio, meaning 'seven plough oxen', although he added that in ancient times only two of the stars were considered oxen, the other five forming a wagon. On a star map of 1524 the German astronomer Peter Apian (1495–1552) showed Ursa Major as a team of three horses pulling a four-wheeled cart, which he called Plaustrum, harking back to the Roman tradition. The word septentrional was commonly used in Latin as a synonym for 'north'.

In mythology, the great bear is identified with two separate characters: Callisto, a paramour of Zeus; and Adrasteia, one of the ash-tree nymphs who nursed the infant Zeus. To complicate matters, there are several different versions of each story, particularly the one involving Callisto.

The story of Callisto

Callisto is usually said to have been the daughter of Lycaon, king of Arcadia in the central Peloponnese. (An alternative story says that she is not Lycaon's

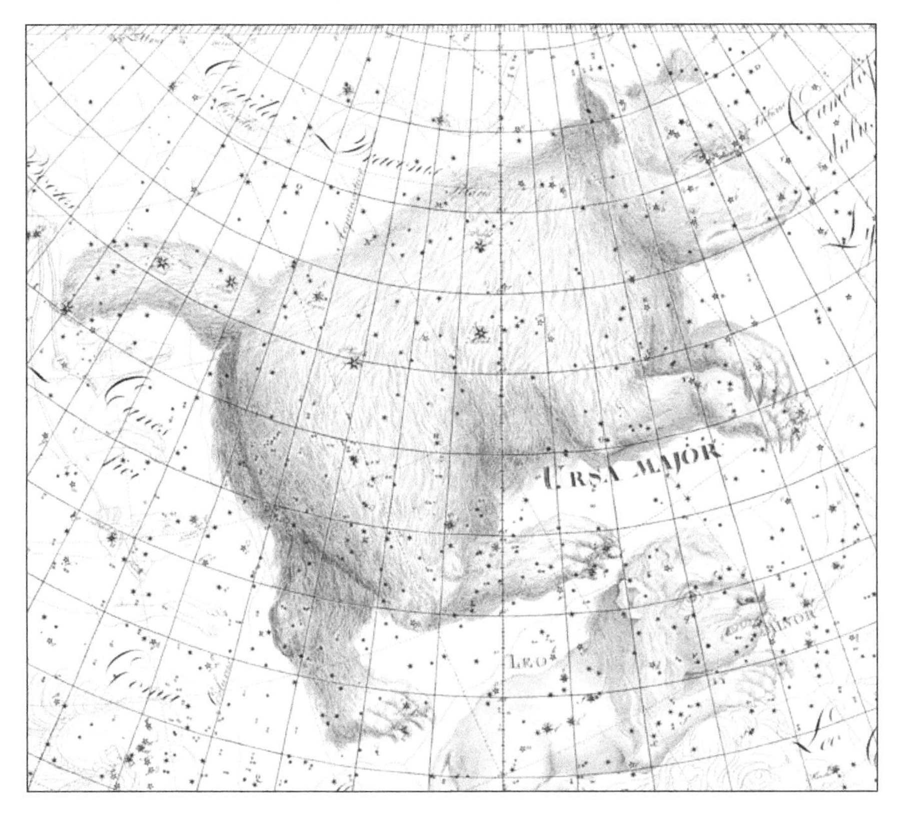

Ursa Major as depicted on Chart VI of the *Uranographia* of Johann Bode (1801). The familiar shape popularly known as the Plough or Big Dipper is made up of seven stars in the body and tail of the bear. (Deutsches Museum)

daughter but the daughter of Lycaon's son Ceteus. In this version, Ceteus is identified with the constellation Hercules, kneeling and holding up his hands in supplication to the gods at his daughter's transformation into a bear.)

Callisto joined the retinue of Artemis, goddess of hunting. She dressed in the same way as Artemis, tying her hair with a white ribbon and pinning together her tunic with a brooch, and she soon became the favourite hunting partner of Artemis, to whom she swore a vow of chastity. One afternoon, as Callisto laid down her bow and rested in a shady forest grove, Zeus caught sight of her and was entranced. What happened next is described fully by Ovid in Book II of his *Metamorphoses*. Cunningly assuming the appearance of Artemis, Zeus entered the grove to be greeted warmly by the unsuspecting Callisto. He lay beside her and embraced her. Before the startled girl could react, Zeus revealed his true self and, despite Callisto's struggles, had his way with her. Zeus returned to Olympus, leaving the shame-filled Callisto scarcely able to face Artemis and the other nymphs.

On a hot afternoon some months later, the hunting party came to a cool river and decided to bathe. Artemis stripped off and led them in, but Callisto hung back. As she reluctantly undressed, her advancing pregnancy was finally revealed. She had broken her vow of chastity! Artemis, scandalized, banished Callisto from her sight.

Callisto becomes a bear

Worse was to come when Callisto gave birth to a son, Arcas. Hera, the wife of Zeus, had not been slow to realize her husband's infidelity and was now determined to take revenge on her rival. Hurling insults, Hera grabbed Callisto by her hair and pulled her to the ground. As Callisto lay spreadeagled, dark hairs began to sprout from her arms and legs, her hands and feet turned into claws and her beautiful mouth which Zeus had kissed turned into gaping jaws that uttered growls.

For 15 years Callisto roamed the woods in the shape of a bear, but still with a human mind. Once a huntress herself, she was now pursued by hunters. One day she came face to face with her son Arcas. Callisto recognized Arcas and tried to approach him, but he backed away in fear. He would have speared the bear, not knowing it was really his mother, had not Zeus intervened by sending a whirlwind that carried them up into heaven, where Zeus transformed Callisto into the constellation Ursa Major and Arcas into Boötes.

Hera was now even more enraged to find her rival glorified among the stars, so she consulted her foster parents Tethys and Oceanus, gods of the sea, and persuaded them never to let the bear bathe in the northern waters. Hence, as seen from mid-northern latitudes, the bear never sets below the horizon.

That this is the most familiar version of the myth is due to Ovid's preeminence as a storyteller, but there are other versions, some older than Ovid. Eratosthenes, for instance, says that Callisto was changed into a bear not by Hera but by Artemis as a punishment for breaking her vow of chastity. Later, Callisto the bear and her son Arcas were captured in the woods by shepherds who took them as a gift to King Lycaon. Callisto and Arcas sought refuge in the temple of Zeus, unaware that Arcadian law laid down the death penalty for trespassers. (Yet another variant says that Arcas chased the bear into the temple while hunting – see Boötes, page 53.) To save them, Zeus snatched them up and placed them in the sky.

The Greek mythographer Apollodorus says that Callisto was turned into a bear by Zeus to disguise her from his wife Hera. But Hera saw through the ruse and pointed out the bear to Artemis who shot her down, thinking that she was a wild animal. Zeus sorrowfully placed the image of the bear in the sky.

Other identifications

Aratus makes a completely different identification of Ursa Major. He says that the bear represents one of the nymphs who raised Zeus in the cave of Dicte on Crete. That cave, incidentally, is a real place where local people still proudly point out the supposed place of Zeus's birth. Rhea, his mother, had smuggled Zeus to Crete to escape Cronus, his father. Cronus had swallowed all his previous children at birth for fear that one day they would overthrow him – as Zeus eventually did. Apollodorus names the nurses of Zeus as Adrasteia and

Ida, although other sources give different names. Ida is represented by the neighbouring constellation of Ursa Minor, the little bear.

These nymphs looked after Zeus for a year, while armed Cretan warriors called the Curetes guarded the cave, clashing their spears against their shields to drown the baby's cries from the ears of Cronus. Adrasteia laid the infant Zeus in a cradle of gold and made for him a golden ball that left a fiery trail like a meteor when thrown into the air. Zeus drank the milk of the she-goat Amaltheia with his foster-brother Pan. Zeus later placed Amaltheia in the sky as the star Capella, while Adrasteia became the great bear – although why Zeus turned her into a bear is not explained.

An enduring puzzle concerning Ursa Major and its companion Ursa Minor is why they came to be regarded as bears when they do not look at all bear-like. Both of the celestial bears have long tails, which real bears do not, an anatomical oddity which the mythologists never explained. Thomas Hood, an English astronomical writer of the late 16th century, offered the tongue-in-cheek suggestion that the tails had become stretched when Zeus pulled the bears up into heaven. 'Other reason know I none', he added apologetically.

Stars of Ursa Major

Two stars in the bowl of the Dipper, Dubhe and Merak (Alpha and Beta Ursae Majoris), are popularly termed the Pointers because a line drawn through them points to the north celestial pole. Dubhe's name comes from the Arabic *al-dubb*, 'the bear', while Merak comes from the Arabic word *al-maraqq* meaning 'the flank' or 'groin'. At the tip of the bear's tail lies Eta Ursae Majoris, known as Alkaid, from the Arabic *al-qa'id* meaning 'the leader'. An alternative name was Benetnasch, from the Arabic *banat na'sh* meaning 'daughters of the bier', for the Arabs regarded this figure not as a bear but as a bier or coffin. They saw the tail of the bear as a line of mourners (the 'daughters') leading the coffin.

Second in line along the tail is the wide double star Zeta Ursae Majoris. The two members of the double, visible separately with keen eyesight, are called Mizar and Alcor. They were depicted as a horse and its rider on the 1524 star chart of Peter Apian, apparently following a popular German tradition. The name Mizar is a corruption of the Arabic *al-maraqq*, the same origin as the name Merak. Its companion, Alcor, gets its name from a corruption of the Arabic *al-jaun*, meaning 'the black horse or bull'. This is the same origin as the name Alioth which is applied to the next star along the tail, Epsilon Ursae Majoris.

The name the Arabs used for Alcor was *al-suha*, meaning the overlooked or neglected one. (Ptolemy certainly overlooked it, as he did not include it in the *Almagest*.) The 10th-century Arabic astronomer al-Ṣūfī noted that this star was used by people to test their eyesight. He quoted an Arabic saying: 'I showed him *al-suha* and he showed me the Moon', as a comparison between people with good and bad eyesight. Delta Ursae Majoris is named Megrez, from the Arabic word *maghriz* meaning 'root of the tail'. Gamma Ursae Majoris is called Phecda, from the Arabic word *fakhidh* meaning 'the thigh'.

In addition to the famous seven stars of the Plough or Dipper there are three pairs of stars that mark the feet of the bear. The Arabs imagined these as forming the hoof prints of a leaping gazelle; according to an Arabic folk tale, the gazelle jumped when it heard the lion hit its tail on the ground. The pair Nu and Xi Ursae Majoris (marking the bear's right hind paw according to Ptolemy) are called Alula Borealis and Alula Australis. The word Alula comes from an Arabic phrase meaning 'first leap'; the distinctions 'northern' (Borealis) and 'southern' (Australis) are added in Latin. The second leap is represented by Lambda and Mu Ursae Majoris, known as Tania Borealis and Tania Australis; these stars were described by Ptolemy as being in the bear's left hind paw. The third leap (and the bear's front left paw) is represented by Iota and Kappa Ursae Majoris; Iota has the name Talitha, from the Arabic meaning 'third', but Kappa is unnamed.

In the *Almagest*, Ptolemy listed 27 stars as members of the constellation, plus another eight that he regarded as lying outside it (i.e. they were unformed). Two of these unformed stars were later incorporated by Johannes Hevelius into one of his new constellations, Canes Venatici, where they are now known as Alpha and Beta CVn. The others became part of another Hevelius invention, Lynx.

Ursa Minor

The little bear

Genitive: Ursae Minoris
Abbreviation: UMi
Size ranking: 56th
Origin: One of the 48 Greek constellations listed by Ptolemy in the *Almagest*
Greek name: Ἄρκτος Μικρά (Arktos Mikra)

The little bear was said by the Greeks to have been first named by the astronomer Thales of Miletus, who lived from about 625 to 545 BC. The earliest reference to it seems to have been made by the poet Callimachus of the third century BC, who reported that Thales 'measured out the little stars of the Wain [wagon] by which the Phoenicians sail'. The little bear was evidently unknown to Homer, two centuries before Thales, for he wrote only of the great bear, never mentioning its smaller counterpart.

It is not clear whether Thales actually invented the constellation or merely introduced it to the Greeks, for Thales was reputedly descended from a Phoenician family and, as Callimachus said, the Phoenicians navigated by reference to Ursa Minor rather than Ursa Major. Aratus points out that although the little bear is smaller and fainter than the great bear, it lies closer to the pole and hence provides a better guide to true north. We have the word of Eratosthenes that the Greeks also knew Ursa Minor as Φοινίκη (Phoenike), i.e. the Phoenician. In the sky the two bears stand back to back, facing in opposite directions, with the tail of Draco the dragon between them.

Aratus called the constellation Κυνόσουρα (Kynosoura, or Cynosura in Latin transliteration), Greek for 'dog's tail'. This is the origin of the English word

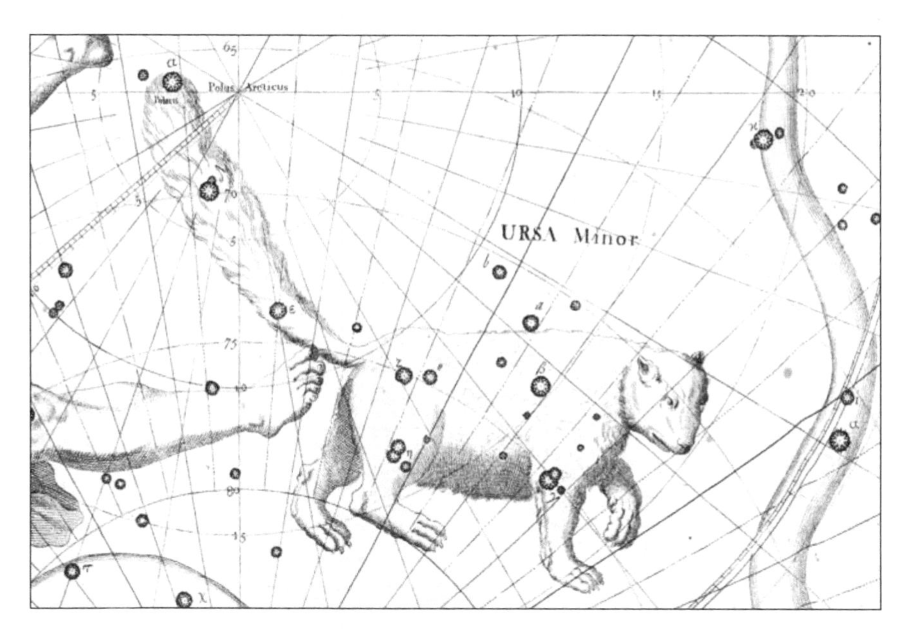

Ursa Minor from the *Atlas Coelestis* of John Flamsteed (1729). Polaris, the north pole star, lies at the tip of its unnaturally long tail. (University of Michigan Library)

cynosure, meaning 'guiding star'. According to Aratus the little bear represents one of the two nymphs who nursed the infant Zeus in the cave of Dicte on Crete. Apollodorus tells us that the nurses' names were Adrasteia and Ida. Ursa Minor commemorates Ida while Adrasteia, the senior of the two, is Ursa Major.

In the *Almagest* it appears under the Greek name Ἄρκτος Μικρά (Arktos Mikra). Ptolemy catalogued seven stars as part of the little bear, four in the body and three in the tail; an eighth star (the modern-day 5 UMi) was also listed but was regarded as lying outside the constellation. These seven stars have a similar ladle shape to the Big Dipper of Ursa Major, and so are popularly termed the Little Dipper. At the end of the little bear's tail (or, alternatuvely, the dipper's handle) is the star Alpha Ursae Minoris, commonly known by the Latin name Polaris because it is currently the nearest bright star to the north celestial pole, although that has not always been the case.

Polaris and the north celestial pole

In Ptolemy's day (2nd century AD) there was no bright star near the north celestial pole. The star we know as Polaris was then some 11° away. Kochab (Beta Ursae Minoris) was closer, but only by a couple of degrees. Over the ensuing centuries, though, the effect of precession slowly moved the celestial pole towards Alpha Ursae Minoris.

Contrary to common belief, Polaris is not particularly bright. It is in fact of magnitude 2.0, just on the fringe of the top 50 brightest stars as seen by the naked eye. Currently it lies within a degree of the exact north celestial pole, close enough to make it an excellent guide star for navigators. Polaris will reach its closest to the north celestial pole around AD 2100, when the separation will be less than half a degree. After that, precession will just as inexorably move the celestial pole away from it again.

Other stars of Ursa Minor

The second star in the little bear's tail, Delta Ursae Minoris, is called Yildun, a mis-spelling of the Turkish word *yildiz* meaning 'star'. According to the star name expert Paul Kunitzsch this was wrongly thought to be a Turkish name for the pole star in Renaissance times, and it has since been arbitrarily applied to the star nearest to the true pole star.

Two stars in the bowl of the Little Dipper, Beta and Gamma Ursae Minoris, are sometimes referred to as the Guards or Guardians of the Pole. Their names are Kochab and Pherkad, and they were seen by the Arabs as a pair of calves. Paul Kunitzsch has been unable to trace the origin of Kochab with certainty, but thinks that it may come from the Arabic word *kaukab* meaning 'star'. Pherkad is from *al-farkadan*, meaning 'the two calves'.

Vela

The sails

Genitive: Velorum
Abbreviation: Vel
Size ranking: 32nd
Origin: Part of the original Greek constellation Argo Navis

Vela is one of the three sections into which the French astronomer Nicolas Louis de Lacaille divided the oversized Greek constellation of Argo Navis, the ship of the Argonauts, in his southern star catalogue of 1756. In that preliminary catalogue he gave it the French name Voilure du Navire, which was Latinized to Vela in his final catalogue of 1763, *Coelum Australe Stelliferum*.

Vela represents the ship's sails; the other sections are Carina, the keel, and Puppis, the stern. Lacaille wrote: 'I have called the sails [i.e. Vela] everything outside the vessel between the edges and the horizontal mast, or the spar on which the sail is reefed'. Not only did Lacaille dismantle Argo Navis, he also relabelled its stars, since he was dissatisfied with Bayer's earlier arrangement. But he still used only one set of Greek letters for all three parts of Argo. As a result, Vela possesses no stars labelled Alpha or Beta, since these letters were allocated to the two brightest stars in Carina. For fainter stars in each constellation Lacaille turned to Roman letters, both lowercase and uppercase.

Vela's brightest star is Gamma Velorum, a second-magnitude double star. Delta and Kappa Velorum, along with Epsilon and Iota Carinae in adjoining Carina, form a cruciform shape known as the False Cross; this is sometimes mistaken for the true Southern Cross, although it is larger and fainter than the real thing. For the full story of Argo, and an illustration, see page 187.

Virgo
The virgin

Genitive: Virginis
Abbreviation: Vir
Size ranking: 2nd
Origin: One of the 48 Greek constellations listed by Ptolemy in the *Almagest*
Greek name: Παρθένος (Parthenos)

Virgo is the second-largest constellation in the sky, exceeded only by the much fainter Hydra. The Greeks including Ptolemy called the constellation Παρθένος (Parthenos). She is usually identified as Dike, goddess of justice, who was the daughter of Zeus and Themis; but she is also known as Astraeia, daughter of Astraeus (father of the stars) and Eos (goddess of the dawn). Virgo is depicted with wings, reminiscent of an angel, holding an ear of wheat in her left hand (the star Spica). Dike features as the impartial observer in a moral tale depicting mankind's declining standards. It was a favourite tale of Greek and Roman mythologists, and its themes still sound familiar today.

Dike was supposed to have lived on Earth in the Golden Age of mankind, when Cronus ruled Olympus. It was a time of peace and happiness, a season

Virgo depicted in the *Atlas Coelestis* of John Flamsteed (1729). In her right hand she carries a palm frond, while in her left hand she holds an ear of wheat marked by the bright star Spica. (University of Michigan Library)

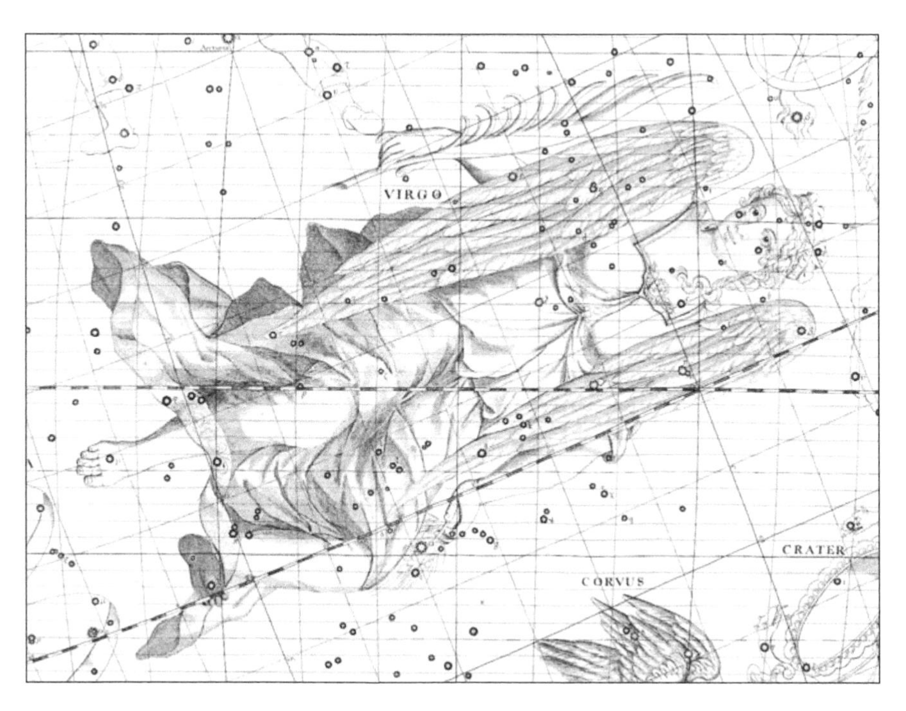

of perennial spring when food grew without cultivation and humans never grew old. Men lived like the gods, not knowing work, sorrow, crime, or war. Dike moved among them, dispensing wisdom and justice.

Then, when Zeus overthrew his father Cronus on Olympus, the Silver Age began, inferior to the age that had just passed. In the Silver Age, Zeus shortened springtime and introduced the yearly cycle of seasons. Humans in this age became quarrelsome and ceased to honour the gods. Dike longed for the idyllic days gone by. She assembled the human race and spoke sternly to them for forsaking the ideals of their ancestors. 'Worse is to come', she warned them. Then she spread her wings and took refuge in the mountains, turning her back on mankind.

Finally came the Ages of Bronze and Iron, when humans descended into violence, theft, and war. Unable to endure the sins of humanity any longer, Dike abandoned the Earth and flew up to heaven, where she sits to this day next to the constellation of Libra, which some see as the scales of justice.

Other identifications

There are other goddesses who can claim identity with Virgo. One is Demeter, the corn goddess, who was daughter of Cronus and Rhea. By her brother Zeus she had a daughter, Persephone (also called Kore, meaning 'maiden'). Persephone might have remained a virgin for ever had not her uncle, Hades, god of the Underworld, kidnapped her while she was out picking flowers one day at Henna in Sicily. Hades swept her aboard his chariot drawn by four black horses and galloped with her into his underground kingdom, where she became his reluctant queen.

Demeter, having scoured the Earth for her missing daughter without success, cursed the fields of Sicily so that the crops failed. In desperation she asked the Great Bear what he had seen, for he never sets, but since the abduction had taken place during the day he referred her to the Sun, who finally told her the unwelcome truth.

Demeter angrily confronted Zeus, father of Persephone, and demanded that he order his brother Hades to return the girl. Zeus agreed to try; but already it was too late, because Persephone had eaten some pomegranate seeds while in the Underworld and, once having done that, she could never return permanently to the land of the living. A compromise was reached in which Persephone would spend half (some say one-third) of the year in the Underworld with her husband, and the rest of the year above ground with her mother. Clearly, this is an allegory on the changing seasons.

Eratosthenes offers the additional suggestion that Virgo might be Atargatis, the Syrian fertility goddess, who was sometimes depicted holding an ear of corn. But this seems to be a mistake because Atargatis is identified with the constellation Piscis Austrinus. Hyginus, more plausibly, equates Virgo with Erigone, the daughter of Icarius, who hanged herself after the death of her father. In this story, Icarius became the constellation Boötes, which adjoins Virgo to the north, and Icarius's dog Maera became the star Procyon (see Boötes, page 54, and Canis Minor, page 64).

Eratosthenes and Hyginus both name Tyche, the goddess of fortune, as another identification of Virgo; but Tyche was usually represented holding the horn of plenty (cornucopia) rather than an ear of grain. In the sky, the ear of corn is represented by the first-magnitude star Spica, a Latin name meaning 'ear of grain'. The name in Greek, Στάχυς (Stachys), has the same meaning.

Beta Virginis is called Zavijava, from an Arabic name meaning 'the angle'; in the *Almagest*, Ptolemy located this star on the top of Virgo's left wing. Gamma Virginis, also in the left wing, is called Porrima, after a Roman goddess. According to Ovid in his *Fasti*, Porrima and her sister Postverta were the sisters or companions of the prophetess Carmenta. Porrima sang of events in the past, while Postverta sang of what was to come.

Epsilon Virginis, on Virgo's right wing, is named Vindemiatrix, from the Latin meaning 'grape-gatherer' or 'vintager', because its first visible rising before the Sun in August marked the beginning of each year's vintage. Ovid in his *Fasti* tells us that this star commemorates a boy named Ampelus (the Greek word for 'vine') who was loved by Dionysus, god of wine. While picking grapes from a vine that trailed up an elm tree, Ampelus fell from a branch and was killed; Dionysus placed him among the stars.

This star's original Greek name, Προτρυγητήρ (Protrygeter), also means 'grape gatherer', the same as in Latin. Its importance as a calendar star is demonstrated by the fact that it was one of the few stars named by Aratus and, at third magnitude, was far fainter than the others.

Virgo, incidentally, contains the autumnal equinox, the point at which the Sun crosses the celestial equator heading south; this occurs on September 22 or 23 each year. In ancient times the autumnal equinox lay in Libra, and hence it is still sometimes referred to as 'the first point of Libra'. However, because of the effect of precession, the autumnal equinox crossed the modern constellation boundary from Libra into Virgo around 730 BC. It continues to move, and will eventually reach Leo in AD 2439.

Volans

The flying fish

Genitive: Volantis
Abbreviation: Vol
Size ranking: 76th
Origin: The 12 southern constellations of Keyser and de Houtman

One of the 12 new constellations introduced at the end of the 16th century by the Dutch navigators Pieter Dirkszoon Keyser and Frederick de Houtman. Volans represents a real type of fish found in tropical waters that can leap out of the water and glide through the air on wings. Sometimes the fish landed on the decks of ships and were used for food. In the sky the flying fish is imagined being chased by the predatory Dorado (page 90), as happens in reality. The

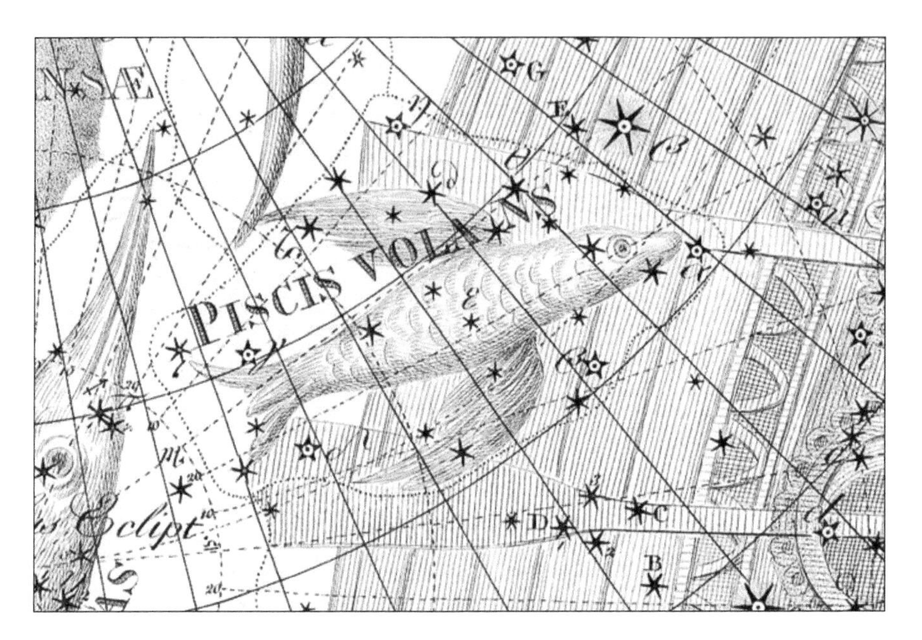

Volans, under its original name Piscis Volans, shown leaping against the side of the ship Argo on Chart XX of the *Uranographia* of Johann Bode (1801). It is chased by the predatory Dorado, at left, which Bode named Xiphias. (Deutsches Museum)

constellation was first depicted in 1598 on a globe by the Dutch cartographer Petrus Plancius under the name Vliegendenvis.

Bayer in 1603 called it Piscis Volans, the Latin title by which it became generally known until the mid-19th century. In 1844 the English astronomer John Herschel proposed shortening it to just Volans. Francis Baily adopted this suggestion in his *British Association Catalogue* of 1845, and it has been known as Volans ever since. Its brightest stars are of only fourth magnitude, none of them are named, and there are no legends associated with it.

Vulpecula

The fox

Genitive: Vulpeculae
Abbreviation: Vul
Size ranking: 55th
Origin: The seven constellations of Johannes Hevelius

A constellation introduced in 1687 by the Polish astronomer Johannes Hevelius, who depicted it as a double figure of a fox, Vulpecula, carrying in its jaws a goose, Anser. Since then the goose has flown (or been eaten), leaving just the fox. Hevelius placed the fox near two other hunting animals, the eagle (the constellation Aquila) and the vulture (which was an alternative identification for Lyra). He explained that the fox was taking the goose to neighbouring Cerberus, another of his inventions – although this part of the tableau has been spoilt, as Cerberus is now obsolete (page 190). The stars used by Hevelius to form Vulpecula had previously been part of a short-lived constellation called Tigris, invented earlier in the 17th century by the Dutch astronomer Petrus Plancius (page 207).

Hevelius himself was somewhat inconsistent in his naming of this constellation. In his star catalogue he named the pair 'Vulpecula cum Ansere', the fox with goose, but showed them separately as 'Anser' and 'Vulpecula' on his *Firmamentum Sobiescianum* star atlas. Others preferred the slightly amended title fox and goose, which is also a traditional pub name in Britain.

Vulpecula contains no named stars and has no legends. Its brightest star, Alpha Vulpeculae of magnitude 4.4, is the only one bearing a Greek letter, allocated by Francis Baily in his *British Association Catalogue* of 1845. Vulpecula is notable for the Dumbbell Nebula, reputedly the most conspicuous of the class of so-called planetary nebulae. The Dumbbell Nebula consists of gas thrown off from a dying star; it takes its name from the double-lobed structure, like a bar-bell, as seen on long-exposure photographs.

On the border with Sagitta is an asterism known as Brocchi's Cluster, or more popularly the Coathanger because of its distinctive bar-and-hook shape. It consists of ten stars of 5th magnitude and fainter and is just visible to the naked eye under good conditions; it was first mentioned by the Arab astronomer al-Ṣūfī in his *Book of the Fixed Stars*, written in AD 964.

The fox and the goose shown as 'Vulpec. & Anser' on the *Atlas Coelestis* of John Flamsteed (1729). The constellation lies between the head of Cygnus, the swan, and Sagitta, the arrow. (University of Michigan Library)

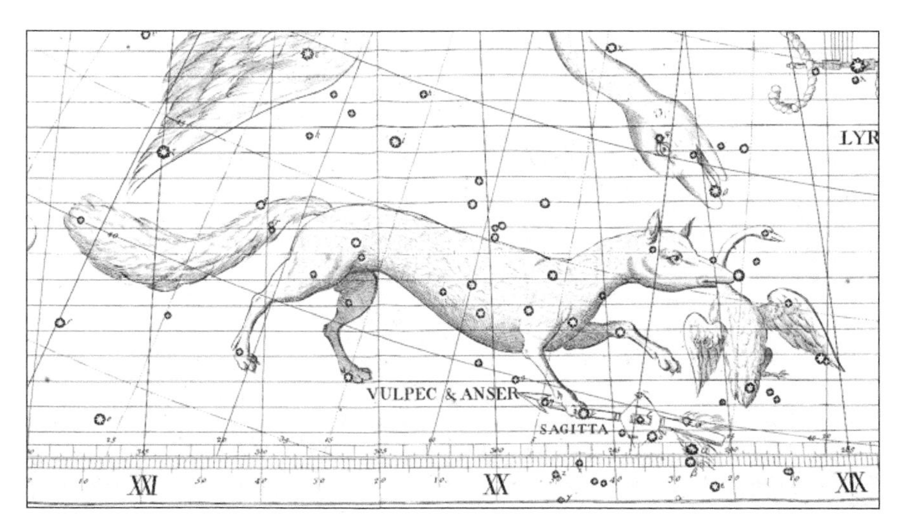

Milky Way

The Milky Way is not, of course, a constellation, but a band of faint light crossing the sky, consisting of countless distant stars. Aratus wrote of the sky being 'cleft all around by a broad band' which he called Γάλα (Gala, the Greek for milk). Eratosthenes called it Κύκλος Γαλαξίας (Kýklos Galaxías), the circle of milk. Ptolemy in the *Almagest* spelled the name γαλακτίας (galaktías).

The Roman writer Manilius compared the Milky Way to the luminous wake of a ship. Ovid in his *Metamorphoses* described it as a road lined on either side by the houses of distinguished gods – 'the Palatine district of high heaven', he termed it. Along this road the gods supposedly travelled to the palace of Zeus.

Eratosthenes tells us that the Milky Way was the result of a trick played by Zeus on his wife Hera so that she would suckle his illegitimate son Heracles and hence make him immortal. Hermes laid the infant Heracles at Hera's breast while she was asleep, but when she woke and realized who the baby was – perhaps by the strength with which he sucked – she pushed him away and her milk squirted across the sky to form the Milky Way.

Manilius listed various explanations for the Milky Way that were current in his day, both scientific and mythological. One suggestion was that it is the seam where the two halves of the heavens are joined – or, conversely, where the two halves are coming apart like a split in the ceiling, letting in light from beyond. Alternatively, said Manilius, it might be a former path of the Sun, now covered in ash where the sky was scorched. Some thought that it could mark the route taken by Phaethon when he careered across the sky in the chariot of the Sun god, Helios, setting the sky on fire (see Eridanus, pages 94–96). Yet again, noted Manilius, it could be a mass of faint stars, an idea attributed to the Greek philosopher Democritus of the fifth century BC, which we now know to be correct. Finally, on a quasi-religious note, Manilius suggested that the Milky Way could be the abode of the souls of heroes who had ascended to heaven.

In Chinese astronomy the Milky Way was *Tianhe*, the Celestial River, also translatable as River of Heaven. Nine stars in Cygnus, including Deneb, represented *Tianjin*, a ford across the river at a point where it appeared to be particularly shallow; the impression of shallowness comes about because a dark cloud of dust in the local spiral arm of our Galaxy obscures part of the Milky Way in this region.

To the Arabs of the Middle Ages the Milky Way was known as *al-madjarra*, from a word meaning a place where something is pulled or drawn along, such as a cart track. It seems that Arab scientists such as al-Bīrūnī (973–1048) understood the Milky Way's true nature as a distant mass of stars, which was finally confirmed by Galileo's telescopic observations in 1610.

CHAPTER FOUR

Obsolete constellations

NON-ASTRONOMERS are often puzzled by the concept of a disused constellation – surely, a constellation is either there or it isn't. However, the patterns we see in the stars are purely a product of human imagination, so humans are free to amend the patterns as they choose – and astronomers did so at will during the heyday of celestial mapping in the 17th and 18th centuries.

The two dozen constellations described in this chapter are a selection of those that, for one reason or another, are no longer recognized by astronomers, although they will be found on old maps. I have included only those constellations that achieved at least some degree of currency, for constellations invented by one astronomer, either in an attempt to make his own name or to flatter his patrons, could be introduced at will and be completely ignored by everyone else. For example, in 1754 the English naturalist John Hill (1714–75) published 15 new constellations in his dictionary of astronomy called *Urania: or, A Compleat View of the Heavens*. These were tucked into spaces between existing figures and represented various unappealing creatures including a toad (Bufo), a leech (Hirudo), a spider (Aranea), an earthworm (Lumbricus), and a slug (Limax). Hill was a noted satirist and he may have been attempting to perpetrate a joke on astronomers – a joke that never caught on.

Several constellations were introduced for mercenary reasons by astronomers wishing to immortalize their kings or governments, usually in the hope that such a gesture would advance their career, as it often did. In 1627 a German astronomer, Julius Schiller (*c.*1580–1627) of Augsburg, attempted to populate the sky entirely with Biblical characters in his atlas called *Coelum Stellatum Christianum*. For example, the familiar constellations of the zodiac were changed to represent the 12 apostles. These attempts to politicize and Christianize the sky were rejected by other astronomers.

Antinous

Antinous (pronounced 'anti-no-us') was the boy lover of the Roman Emperor Hadrian and hence is a real character, not a mythological one, although the story reads like fiction. Antinous was born *c.* AD 110 in the town of Bythinium (also called Claudiopolis), near present-day Bolu in north-western Turkey. At that time the area was a Roman province, and Hadrian is thought to have met Antinous during an official visit. Hadrian, the first openly gay Roman Emperor, was smitten by the boy and groomed him to become his constant companion.

Hadrian's happiness did not last long, though. While on a trip up the Nile in AD 130 Antinous drowned near the present-day town of Mallawi in Egypt. Supposedly an oracle had predicted that the Emperor would be saved from danger by the sacrifice of the object he most loved, and Antinous realized that this description applied to him. Whether the drowning was accident, suicide, or even ritual sacrifice, Hadrian was heartbroken by it. He founded a city called Antinoöpolis near the site of the drowning, declared Antinous a god, and

Antinous carried in the claws of Aquila the eagle, seen on Chart IX of the *Uranographia* of Johann Bode (1801). (Deutsches Museum)

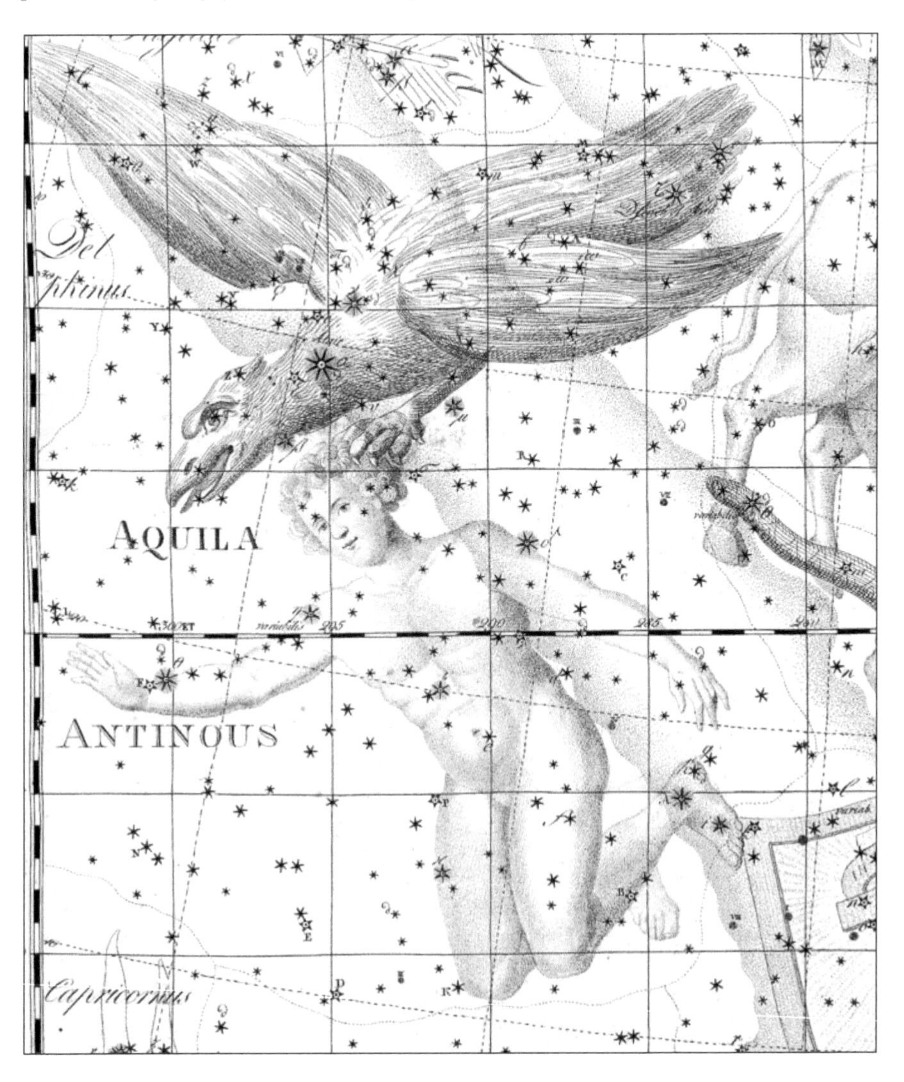

commemorated him in the sky from stars south of Aquila, the eagle, that had not previously been considered part of any constellation.

Antinous was mentioned as a sub-division of Aquila in Ptolemy's *Almagest*, although it is not included among the canonical 48 Greek figures. Ptolemy worked at Alexandria at the mouth of the Nile and he compiled the *Almagest* about 20 years after the famous drowning so he would have known the story; indeed, he might have had a hand in creating the constellation, possibly at Hadrian's request.

The constellation's first known depiction was in 1536 on a celestial globe by the German mathematician and cartographer Caspar Vopel (1511–61); it was shown again in 1551 on a globe by Gerardus Mercator. Tycho Brahe listed it as a separate constellation in his star catalogue of 1602 and it remained widely accepted into the 19th century, when it was eventually remerged with Aquila. Antinous was depicted being carried in the talons of Aquila. Hence he has sometimes been confused with Ganymede, another celestial catamite, who was carried off by an eagle for Zeus.

Argo Navis
The ship Argo

Argo (Ἀργώ in Greek) is a constellation that is not so much disused as dismantled. It was one of the 48 constellations known to Greek astronomers, as listed by Ptolemy in the *Almagest*, but astronomers in the 18th century found it large and unwieldy and so divided it into three parts: Carina, the keel or body; Puppis, the poop (i.e. stern); and Vela, the sails. Were the three parts to be reunited, the resulting figure would be almost 28% larger in area than the current largest constellation, Hydra. The modern constellation Pyxis, the compass, occupies an area next to the mast, but it is not considered a part of the original Argo.

Voyage of the Argo

Argo Navis represents the 50-oared galley in which Jason and his crew of Argonauts sailed to fetch the golden fleece from Colchis, on the eastern shores of the Black Sea in present-day Georgia. This fleece, incidentally, came from the ram which is now represented by the constellation Aries. Jason was the rightful successor to the throne of Iolcus in eastern Greece. But the throne had been seized by his arrogant uncle Pelias while Jason was still a child and there seemed no chance that Jason would inherit it. When Jason had grown into a man, Pelias deceitfully offered to relinquish the throne if Jason could bring back the golden fleece from Colchis. It was a round trip of over 2,000 miles, and Pelias secretly hoped that Jason would perish along the way.

First, he needed a ship capable of such an epic voyage. Jason entrusted its construction to Argus, after whom it was named. Argus built the ship under supervision of the goddess Athene at the port of Pagasae (the modern Volos),

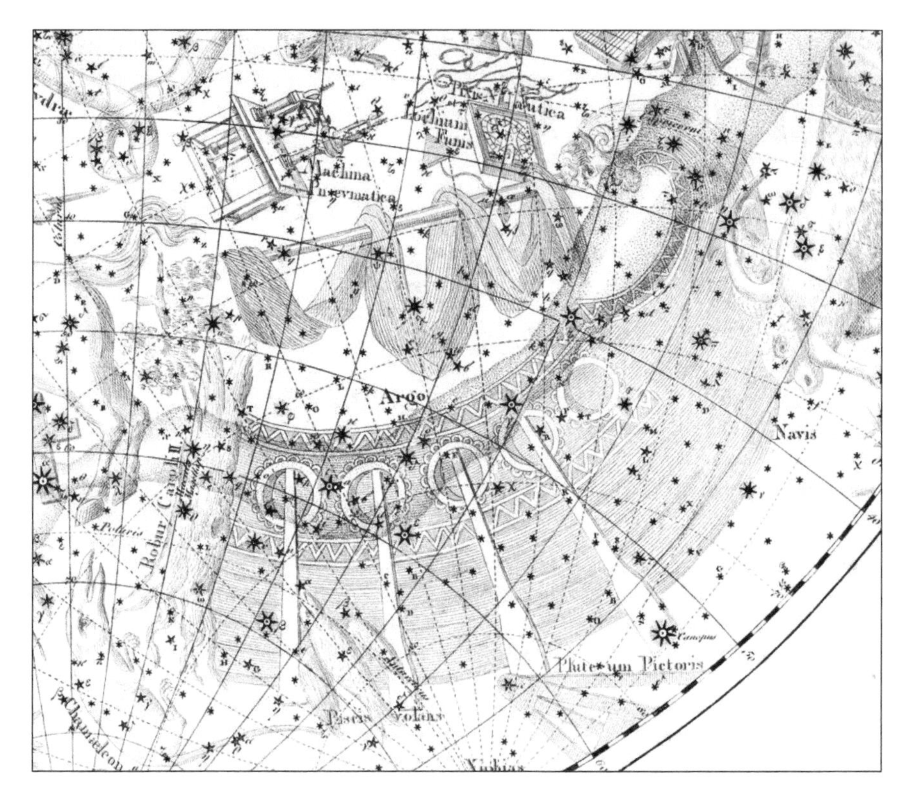

Argo Navis dominates this crowded scene in the southern sky from Chart II of the *Uranographia* of Johann Bode (1801). At lower right lies the bright star Canopus, on the blade of one of the steering oars. At the left, across the prow of the ship, lies Charles's Oak, here called Robur Caroli II, a now-obsolete constellation invented by Edmond Halley (see page 202). The ship's sail is wrapped around a spar that appears to emerge from the stern. Because of Argo's considerable size, cartographers struggled to depict it successfully on a single chart. This attempt by Bode is perhaps the best. (Deutsches Museum)

using timber from nearby Mount Pelion. Into the prow Athene fitted an oak beam from the oracle of Zeus at Dodona in north-western Greece. This area, like the island of Corfu nearby, was once noted for its forests of oak, before later shipbuilders stripped them bare. Being part of an oracle, this oak beam could speak and it was crying out for action by the time the Argo left harbour.

Jason took with him 50 of the greatest Greek heroes, including the twins Castor and Polydeuces, the musician Orpheus, as well as Argus, the ship's builder. Even Heracles interrupted his labours to join the crew.

Apollonius of Rhodes, who wrote the epic story of the ship's voyage to Colchis and back, described Argo as the finest ship that ever braved the sea with oars. Even in the roughest of seas the bolts of Argo held her planks together safely, and she ran as sweetly when the crew were pulling at the oars as she did before the wind. Isaac Newton thought the voyage of the Argo was commemorated in the 12 signs of the zodiac, although the connections are hard to see.

Among the greatest dangers the Argonauts faced *en route* were the Clashing Rocks, or Symplegades, which guarded the entrance to the Black Sea like a pair of sliding doors, crushing ships between them. As the Argonauts rowed along the Bosporus, they could hear the terrifying clash of the Rocks and the thunder of surf. The Argonauts released a dove and watched it fly ahead of them. The Rocks converged on the dove, nipping off its tail feathers, but the bird got through. Then, as the Rocks separated, the Argonauts rowed with all their might. A well-timed push from the divine hand of Athene helped the ship through the Rocks just as they slammed together again, shearing off the mascot from Argo's stern. Argo had become the first ship to run the gauntlet of the Rocks and survive. Thereafter the Clashing Rocks remained rooted apart.

Once safely into the Black Sea, Jason and the Argonauts headed for Colchis. There they stole the golden fleece from King Aeëtes, and made off with it back to Greece by a roundabout route. After their return, Jason left the Argo beached at Corinth, where he dedicated it to Poseidon, the sea god.

Eratosthenes said that the constellation represents the first ocean-going ship ever built, and the Roman writer Manilius concurred. However, this attribution must be wrong because the first ship was actually built by Danaus, father of the 50 Danaids, again with the help of Athene, and he sailed it with his daughters from Libya to Argos.

Argo in the sky

Only the stern of Argo is shown in the sky. Map makers attempted to account for this truncation either by depicting its prow vanishing into a bank of mist, as Aratus described it, or passing between the Clashing Rocks, as shown on Bayer's atlas. Robert Graves recounts the explanation that Jason in his old age returned to Corinth where he sat beneath the rotting hulk of Argo, contemplating past events. Just at that moment the rotten beams of the prow fell off and killed him. Poseidon then placed the rest of the ship among the stars. Hyginus, though, says that Athene placed Argo among the stars from steering oars to sail when the ship was first launched, but says nothing about what happened to the prow.

Argo was first split into three by the French astronomer Nicolas Louis de Lacaille in his catalogue of the southern stars published in 1756 and it now lies permanently dismembered. In the notes to his catalogue Lacaille wrote: 'I have divided [Argo] into three parts, namely la Pouppe [Puppis], le Corps [Carina] & la Voilure [Vela]'. Pyxis, the ship's compass, was also introduced by Lacaille in this same area but he listed it separately, among his 14 new constellations, and so it is not considered a part of Argo.

There is, though, still an echo of Argo's former unity. Lacaille was dissatisfied with Bayer's allocation of Greek letters to the stars of Argo, so he decided to change them. However, when he did so, he used just one sequence of Greek letters, from Alpha to Omega, as though Argo were still a single figure. The designations Alpha and Beta were given to the two brightest stars of Argo, which are in Carina; hence there are no stars labelled Alpha or Beta in either Puppis or Vela. Equally, the brightest star in Puppis is Zeta and the brightest in Vela is Gamma, but there is no Zeta or Gamma Carinae.

Cerberus

Cerberus was the triple-headed monster that guarded the gates of Hades, the realm of the dead, preventing the living from entering and the dead from leaving. For the last and most dangerous of his 12 labours Hercules was sent to the Underworld to capture this fearsome creature (see page 104). He wrestled it into submission with his bare hands and dragged it, writhing and resisting, from the darkness of the Underworld to the unaccustomed brightness of the surface. The constellation commemorating this feat was added to the sky by Johannes Hevelius in his catalogue and atlas of 1687 in which he depicted Cerberus grasped in the outstretched hand of Hercules. Although in mythology Cerberus was described as a three-headed dog, Hevelius and all subsequent map makers illustrated it with three snake heads.

Hevelius's Cerberus replaced another figure, the branch from the tree of the golden apples, that Johann Bayer had previously depicted in the hand of Hercules. In or around 1721 the English cartographer and engraver John Senex (1678–1740), a friend of Edmond Halley, combined Cerberus with the apple branch, Ramus, to produce Cerberus et Ramus. This combined figure first appeared on Senex's chart of the northern celestial hemisphere, *Stellarum Fixarum Hemisphaerium Boreale*. The serpents in this case were wrapped around the apple branch. Johann Bode subsequently showed Cerberus and Ramus together on his *Uranographia* atlas of 1801 as seen below.

Cerberus et Ramus, the three-headed monster wrapped around an apple branch, shown on Chart VIII of the *Uranographia* of Johann Bode (1801). (Deutsches Museum)

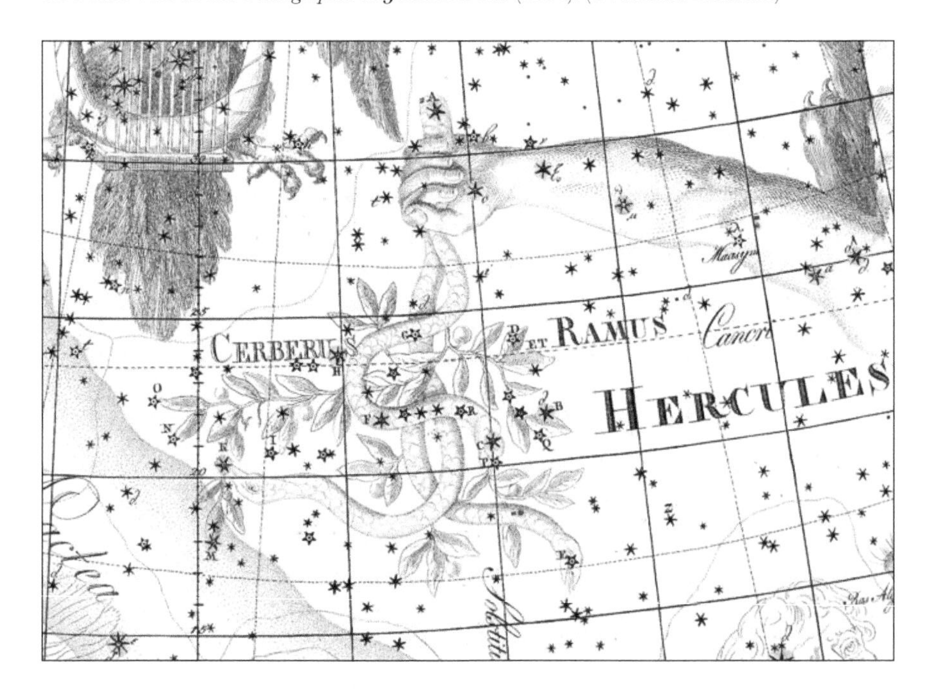

Custos Messium
The harvest keeper

This far-northern constellation was introduced by the French astronomer Joseph Jérôme de Lalande on his celestial globe of 1775. The name Custos Messium is a punning reference to his countryman Charles Messier, the famed comet hunter, and in fact the constellation was often known simply as Messier, particularly in France. Custos Messium consisted of stars of 4th magnitude and fainter between Cepheus and Camelopardalis in what is now northern Cassiopeia. Lalande chose this previously anonymous area of sky because it was here that the comet of 1774 was first seen. The comet was extensively observed by Messier but, ironically, was not discovered by him – the discoverer in this case was actually another Frenchman, Jacques Montaigne.

The British scientist Thomas Young (1773–1829) renamed the figure the Vineyard Keeper on his chart of the northern sky published in 1807, but even this was not enough to broaden its appeal and it withered into obscurity.

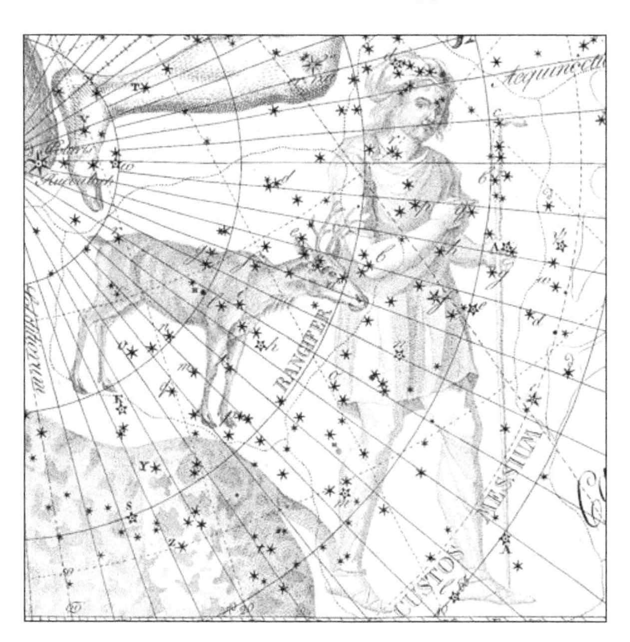

Custos Messsium seen beside Rangifer, the reindeer (page 201), another now-obsolete constellation, on Chart III of the *Uranographia* of Johann Bode (1801). The harvest keeper was depicted as a rustic figure holding a shepherd's crook, equipping him for herding livestock as well as guarding crops. (Deutsches Museum)

Felis
The cat

Cats are often put out for the night, but the French astronomer Joseph Jérôme de Lalande tried to put one into the night sky in the shape of the constellation Felis. Lalande did not himself depict the constellation on any globe or chart but suggested it to Johann Bode, who first showed it on his *Uranographia* atlas of

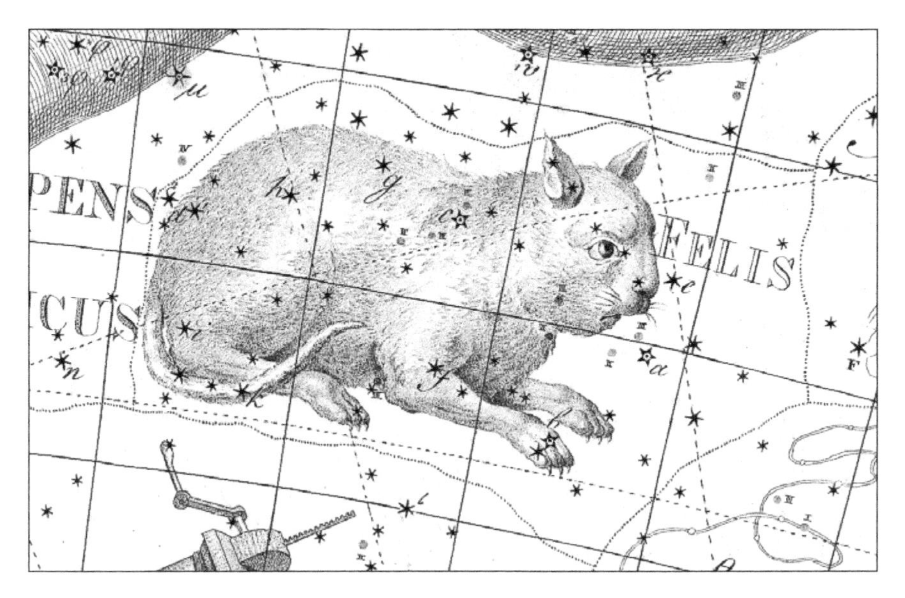

A grumpy-looking Felis on Chart XIX of the *Uranographia* of Johann Bode (1801). The cat crouches under the snaking body of Hydra (top), with Antlia beneath it and Pyxis at lower right. (Deutsches Museum)

1801. Felis consisted of a scattering of faint stars between Hydra to its north and Antlia to the south. Lalande had supplied many star positions to Bode for his atlas and catalogue, including those in this area, and felt that this gave him the right to suggest some new constellations.

Lalande announced his invention in a letter that appeared in *Allgemeine Geographische Ephemeriden*, a German scientific magazine, in 1799. A translation of his imperfect German might be: 'I have inserted between the Ship and the Cup a new constellation, a Cat… I greatly love these animals… I will have it engraved on the charts; the starry sky has tired me enough in my life that I can now have my fun with it.' Other astronomers proved less susceptible to feline charms, though, and the cat eventually vanished into the night.

Gallus

The cockerel

Gallus was formed by the Dutch theologian and cartographer Petrus Plancius and first appeared on his celestial globe of 1612. It lay in the Milky Way south of the celestial equator in the northern part of what is now Puppis. Gallus made its first appearance in print twelve years later on a chart by the German astronomer Jacob Bartsch (*c.*1600–33) in his book *Usus Astronomicus Planisphaerii Stellati* (1624). Bartsch, who was keen to find Biblical references for constellations, said in the accompanying text that Gallus represented the cockerel that crowed after Peter had denied Jesus thrice. Whether this was actually Plancius's intention we cannot tell, for Plancius left no surviving records. Another version of Gallus appeared on a chart published in 1628 by the German astronomer Isaac Habrecht (1589–1633) in his book *Planiglobium Coeleste, et Terrestre.*

Gallus was awkwardly placed in an area that was already occupied by the stern of Argo Navis and so was superfluous. Later astronomers such as Johannes Hevelius returned its stars to Argo and it was soon forgotten.

Globus Aerostaticus

The balloon

A hot-air balloon once floated serenely in the sky south of the zodiacal constellation Capricornus, next to the tail of Piscis Austrinus. It first appeared on the *Uranographia* atlas of Johann Elert Bode in 1801, but it had been suggested to him three years earlier by the French astronomer Joseph Jérôme de Lalande to honour the hot-air balloon invented in the 1780s by the Montgolfier brothers.

In his *Histoire Abrégée de l'Astronomie*, Lalande recalled that his countryman Nicolas Louis de Lacaille had placed instruments of science and arts among the stars of the southern hemisphere and explained: 'I thought the greatest discovery of the French deserved to occupy a place.' At the same time, he and Bode agreed to introduce a constellation representing Gutenberg's printing press; this became Bode's Officina Typographica (page 199). History shows that the printing press has been more influential than the hot-air balloon, but neither remain among the recognized constellations.

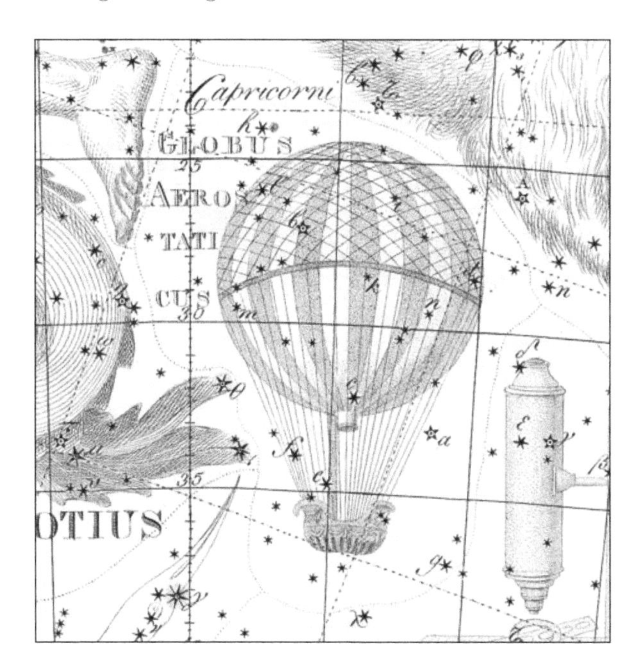

Globus Aerostaticus, rising into the sky with an empty basket, on Chart XVI of the *Uranographia* of Johann Bode (1801). To the left of the balloon is the tail of the southern fish, Piscis Austrinus (called Piscis Notius by Bode), while to the lower right of it is Microscopium. Capricornus is at top. (Deutsches Museum)

Harpa Georgii

George's harp

Maximilian Hell (1720–92), the Hungarian-born director of the Vienna observatory, introduced this constellation in 1789 under the name Psalterium Georgianum, i.e. George's Psaltery, a psaltery being an ancient form of harp. It was intended to honour King George III of England, patron of William Herschel who had discovered the planet Uranus from his back garden in Bath in 1781. This celestial harp lay beneath the front hooves of Taurus the bull and above a bend in the river Eridanus. It first appeared on a chart in a special publication by Hell called *Monumenta, Aere Perenniora, Inter Astra Ponenda.*

Johann Bode changed the name of Hell's new constellation to Harpa Georgii on his *Uranographia* atlas of 1801. Bode depicted it as a more modern form of harp and angled it to fit better between the surrounding constellations (see below). This was the representation that became best-known.

Harpa Georgii comes close to being trampled beneath the left front hoof of Taurus on Chart XII of the *Uranographia* of Johann Bode (1801). The dashed line crossing the top of the harp is the celestial equator. (Deutsches Museum)

Honores Friderici

Frederick's glory

A constellation introduced by Johann Bode in 1787 to commemorate King Frederick the Great of Prussia, who had died the preceding year. Bode originally called it by the German name of Friedrichs Ehre, which can be translated as either 'glory' or 'honour', the former being the more archaic usage, but he Latinized the name to Honores Friderici on his *Uranographia* of 1801. Names for it found on other atlases include Gloria Frederici and Frederici Honores, all with

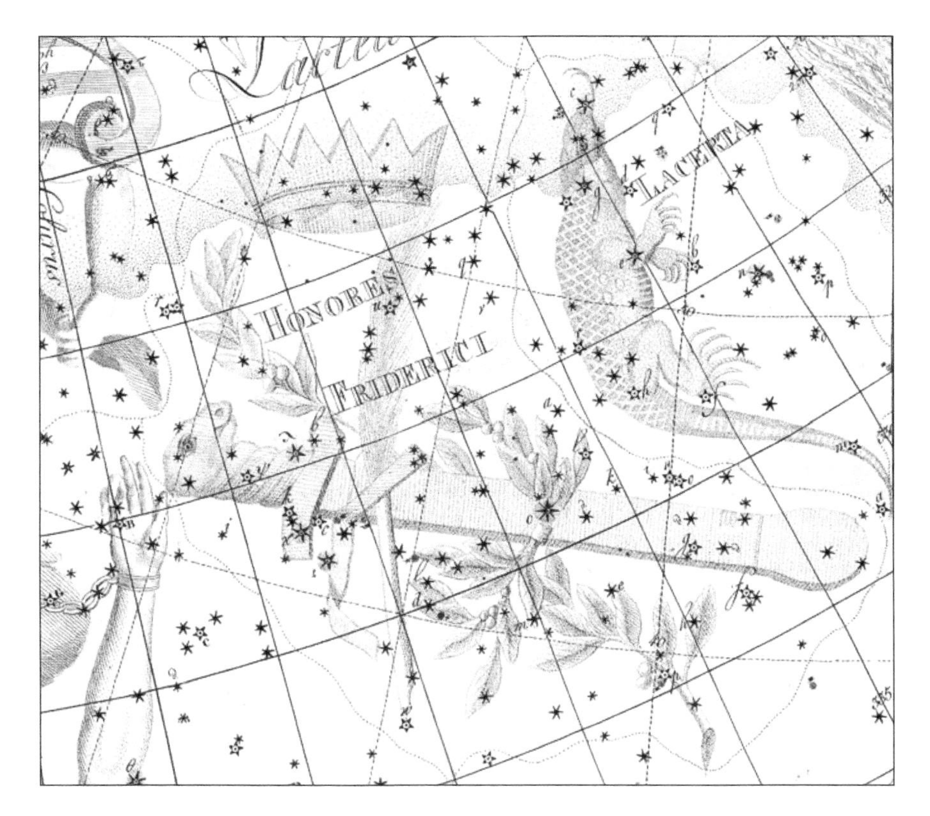

Honores Friderici, between Lacerta the lizard and the hand of Andromeda, as shown on Chart IV of the *Uranographia* of Johann Bode (1801). It consisted of a ceremonial sword entwined with a strand of laurel, a quill pen, and a surmounting crown, all meant to symbolize King Frederick of Prussia as a hero, sage, and peacemaker. (Deutsches Museum)

the same meaning. The constellation was squeezed between the outstretched right arm of Andromeda and the Hevelius invention of Lacerta, the lizard. In this same area the Frenchman Augustin Royer had in 1679 placed his own invention, Sceptrum, representing the French sceptre and hand of justice, commemorating Louis XIV. Neither constellation survived.

Jordanus

Currently there is only one river in the sky, Eridanus, but at one time there were another two: Jordanus and Tigris (for Tigris, see page 207). Jordanus, also known as Jordanus Fluvius or Jordanis, represented the river Jordan and was introduced by the Dutchman Petrus Plancius on his celestial globe of 1612. It rose under the tail of the great bear in what is now the constellation of Canes Venatici, flowed between the bear and Leo (an area now occupied by Leo Minor and Lynx), and ended next to Camelopardalis, another Plancius invention.

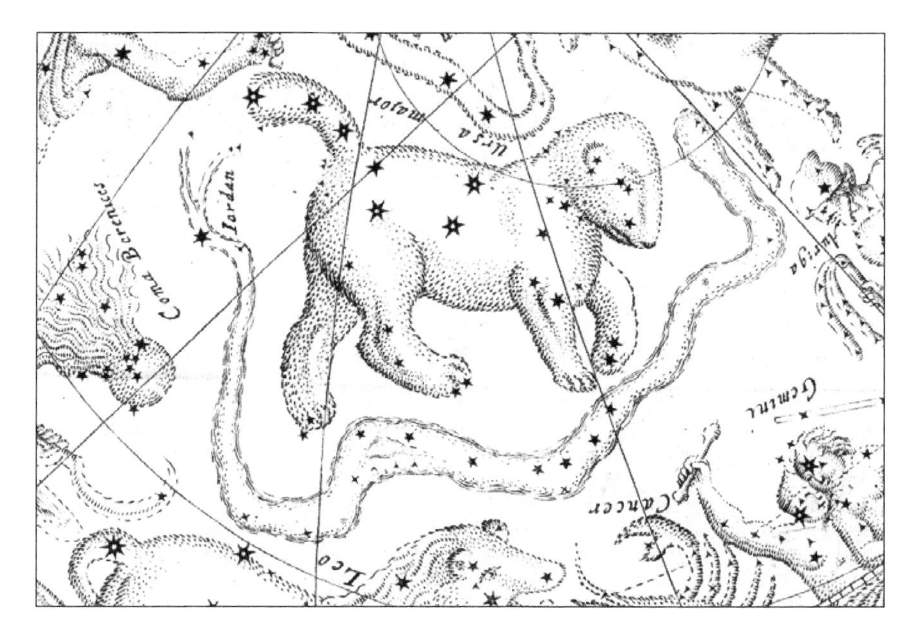

Jordanus flowing between the feet of Ursa Major and the head of Leo as shown in the 1666 edition of Isaac Habrecht's book *Planiglobium Coeleste, et Terrestre*, originally published in 1628. The bright star at the confluence of the twin headwaters at left is the one we now know as Cor Caroli in Canes Venatici. (ECHO, Berlin.)

The German astronomer Jacob Bartsch (*c.*1600–33) was the first to depict the constellation in print, in his book *Usus Astronomicus Planisphaerii Stellati* of 1624, as a result of which he was sometimes erroneously credited with its invention. In that book he explained that the river had two sources, namely Jor and Dan, but not every chart showed it this way, including his own. One that did was by Bartsch's mentor Isaac Habrecht (1589–1633) in 1628 (see illustration above). Jordanus became forgotten during the 18th century once Hevelius had introduced Canes Venatici, Leo Minor, and Lynx in this same area, and it was not shown by Bode.

Lochium Funis

The log and line

A constellation added to the southern sky by Johann Bode in his *Uranographia* atlas of 1801 representing a nautical log and line used for measuring speed and distance travelled at sea. It was coiled around Pyxis, the compass, a previous invention by the Frenchman Nicolas Louis de Lacaille. The device consisted of a weighted piece of flat wood (the log), attached to a long rope with knots tied in it (the line); in operation, the sailors threw the log overboard and counted the number of knots that were paid out in half a minute, as timed by a sand glass.

Bode treated Pyxis and Lochium Funis as a combined figure; on his atlas he enclosed them both within the same constellation boundary and listed their stars together in his accompanying catalogue. Lacaille's Pyxis still exists but Bode's Lochium Funis soon sank without trace. For an illustration, see Pyxis (page 150).

Machina Electrica

The electrical machine

A constellation introduced by the German astronomer Johann Bode on his *Uranographia* atlas of 1801, representing one of the mechanical wonders of the age, an electrostatic generator. Machina Electrica lay in the southern sky between Fornax and Sculptor, two Lacaille constellations.

The device that Bode illustrated in his atlas is of the type devised by the English instrument maker Jesse Ramsden in the late 18th century. It consists of a glass disk turned by a handle (left), from which a cylindrical conductor bar (centre) leads to a Leyden jar to store the charge. Next to the jar is a U-shaped discharge wand, while beneath the conductor bar is a table for experiments with a connecting chain. To make his new figure Bode expropriated stars from both Fornax and Sculptor; worst to suffer was Sculptor, which was halved in extent. Most of the stars, though, were so faint as to be barely visible to the naked eye. The annexed territory has since been returned.

Machina Electrica, shown on Chart XVII of Johann Bode's *Uranographia* star atlas of 1801. Most of the area it occupied was taken from Sculptor, on the right. (Deutsches Museum)

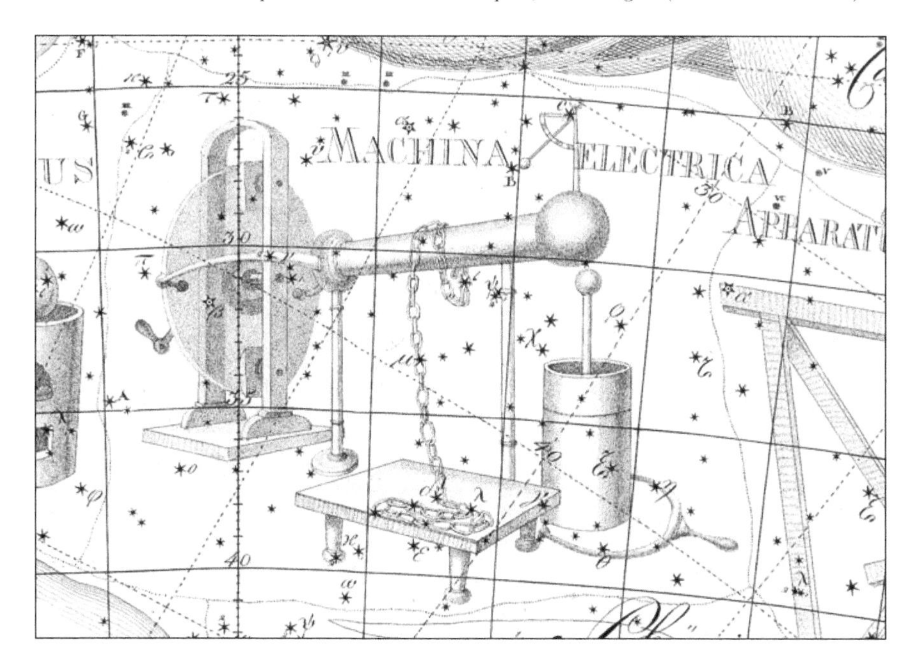

Mons Maenalus

Mount Maenalus

A mountain of Arcadia in the central Peloponnese, introduced as a constellation by Johannes Hevelius in his *Firmamentum Sobiescianum* star atlas of 1687, where it was depicted with Boötes standing on it. It appeared on many later maps, always as part of Boötes, and it never had an independent existence.

The mountain took its name from a character in Greek mythology. Maenalus was said by some mythologists to have been the eldest son of Lycaon, king of Arcadia; this would have made Maenalus brother of Callisto and hence uncle of her son Arcas, whom the constellation Boötes represents. Others, though, say he was actually the son of Arcas and hence the grandson of Callisto. Either way, Maenalus gave his name to the mountain in Arcadia and to the city of Maenalon which he founded. Its modern name is Mainalo.

Mons Maenalus was sacred to the god Pan who frequented it. Ovid in his *Metamorphoses* said that Mons Maenalus bristled with the lairs of wild beasts and was a favourite hunting ground of Diana and her entourage, including Callisto. In saying this, Ovid evidently rejected the story that Maenalus was Callisto's grandson, as the mountain would not yet have got its name.

Mons Maenalus as depicted by Johannes Hevelius on his *Firmamentum Sobiescianum* atlas. The constellation's name was originally engraved on the chart as 'Menalis' but this was subsequently altered to Mænalus. See also Boötes, page 54. (National Library of Poland)

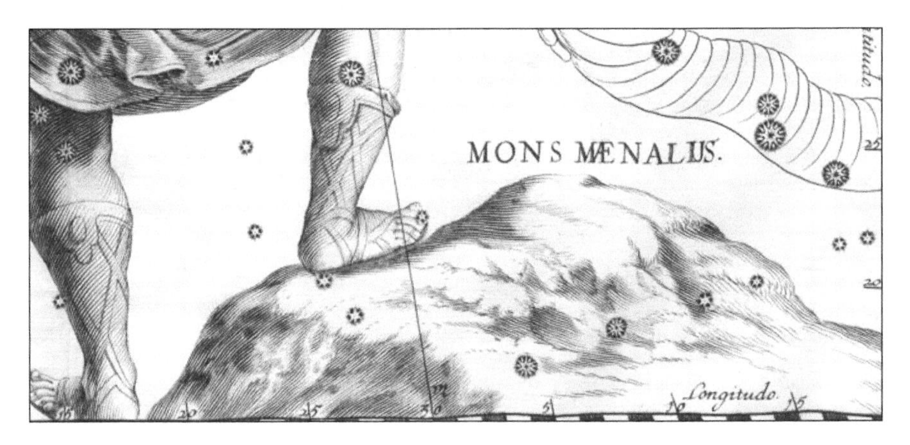

Musca Borealis

The northern fly

This abandoned constellation, which lay in the northern part of present-day Aries, has a confusing history. It was introduced on a globe of 1612 by the Dutchman Petrus Plancius under the name Apes, the Bee. He created it from

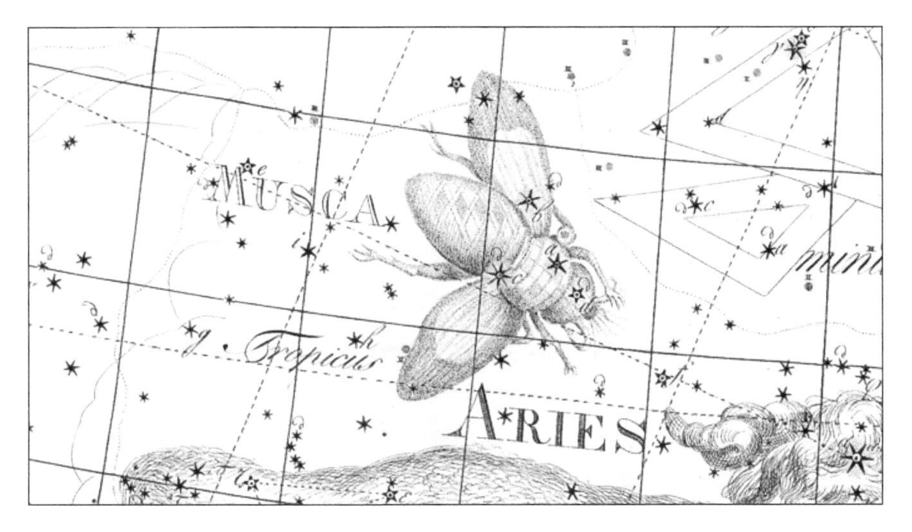

Musca Borealis buzzes above the back of Aries the ram on Chart XI in the *Uranographia* of Johann Bode (1801). (Deutsches Museum)

four of Ptolemy's unformed stars which were described in the *Almagest* as lying 'over the rump' of Aries. Johannes Hevelius changed the type of insect when he renamed it Musca, the fly, on his *Firmamentum Sobiescianum* atlas of 1687.

This was confusing, since there already was a Musca in the southern hemisphere, created almost a century earlier from the observations of Keyser and de Houtman. Johann Bode showed Musca on his *Uranographia* atlas, but within the borders of Aries (see illustration above). The constellation later became known as Musca Borealis, the northern fly, to distinguish it from its southern namesake. This longer name seems to have first appeared on Alexander Jamieson's *Celestial Atlas* of 1822. Eventually, the northern fly was swatted by astronomers, although the southern Musca remains.

To add to the confusion, the same stars were used by the Frenchman Ignace-Gaston Pardies (1636–73) to form Lilium, the fleur-de-lis of France. Lilium appeared in a star atlas by Pardies entitled *Globi coelestis in tabulas planas redacti descriptio* that was published in 1674, the year after his death, but was a very short-lived addition to the sky.

Officina Typographica
The printing shop

Johann Bode introduced this constellation on his *Uranographia* star atlas of 1801 to commemorate invention of the printing press some 350 years earlier by his German compatriot Johannes Gutenberg. It lay in what is now the northern part of Puppis, between Canis Major and the hind legs of Monoceros. As depicted by Bode (see overleaf), Officina Typographica consisted of a case of

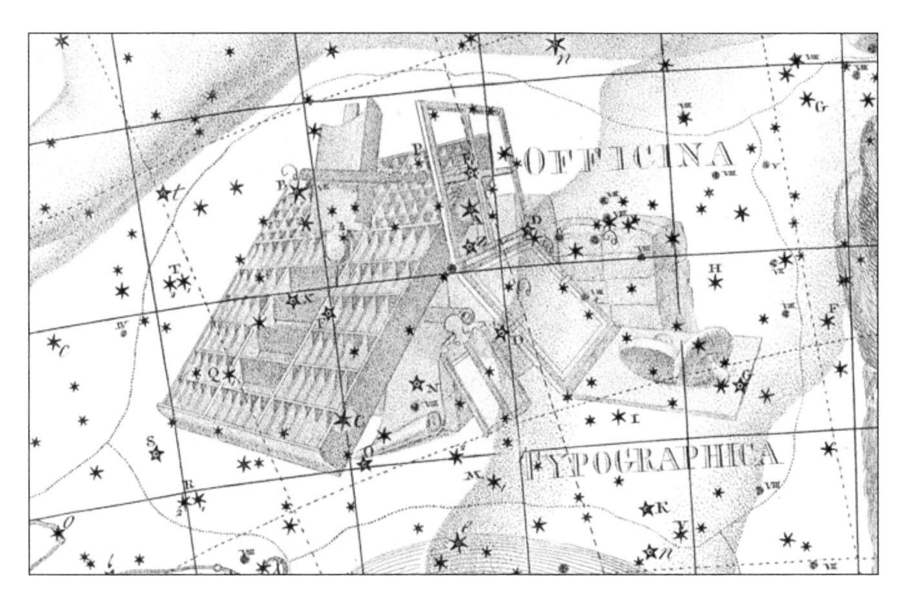

Officina Typographica, depicting a printer's workshop, was invented by Johann Bode and made its first appearance on Chart XVIII of his *Uranographia* atlas of 1801. Part of the hind quarters of Monoceros, the unicorn, can be seen above it. (Deutsches Museum)

movable type with composing stick; the frisket, a frame with four windows that folded over the printing paper; the tympan, on which the paper was placed; two inking pads to ink the type; and stacks of paper in the background. The printing press itself, which pressed the inked type against the paper, was not shown.

According to the French astronomer Joseph Jérôme de Lalande in his *Histoire Abrégée de l'Astronomie*, he and Bode agreed in 1798 to create two new constellations, 'la Presse de Gutemberg [*sic*] et la Globe de Montgolfier', thereby commemorating two great inventions of Germany and France respectively. The Montgolfier balloon became Globus Aerostaticus (see page 193). Both these new Franco-German figures made their debuts on Bode's *Uranographia* in 1801. Although both were subsequently shown on many popular maps they were not eventually accepted, perhaps because the motives for their invention were too overtly nationalistic.

Quadrans Muralis

The mural quadrant

Although now abandoned as a constellation, its name lives on in the annual meteor shower known as the Quadrantids that radiates from this area every January. Quadrans Muralis occupied what is now the northern part of Boötes, near the end of the Plough's handle (or, alternatively, the tip of the great bear's tail). The constellation was invented in the 1790s by the French astronomer

Joseph Jérôme de Lalande, director of the observatory of l'École Militaire in Paris. It commemorated the observatory's wall-mounted quadrant ('muralis' is Latin for 'wall') which he used for measuring star positions. Lalande took his cue from his countryman Nicolas Louis de Lacaille who, forty years earlier, had populated the southern sky with constellations representing scientific instruments. Lalande followed suit by putting a quadrant in the northern sky.

Quadrans Muralis first made its appearance under the name Le Mural in the 1795 edition of the *Atlas Céleste* of Jean Fortin (1750–1831), another Frenchman, which was edited by Lalande and his colleague Pierre Méchain. Johann Bode Latinized its name from Le Mural to Quadrans Muralis in 1801 on his own atlas, *Uranographia*. The constellation was evidently considered well established in the 1830s when the Quadrantid meteor shower was first recognized and named after it, but even this was not enough to save it from extinction by the end of the 19th century.

Quadrans Muralis lay just above the outstretched arm of Boötes, as shown on Chart VII of the *Uranographia* of Johann Bode (1801). (Deutsches Museum)

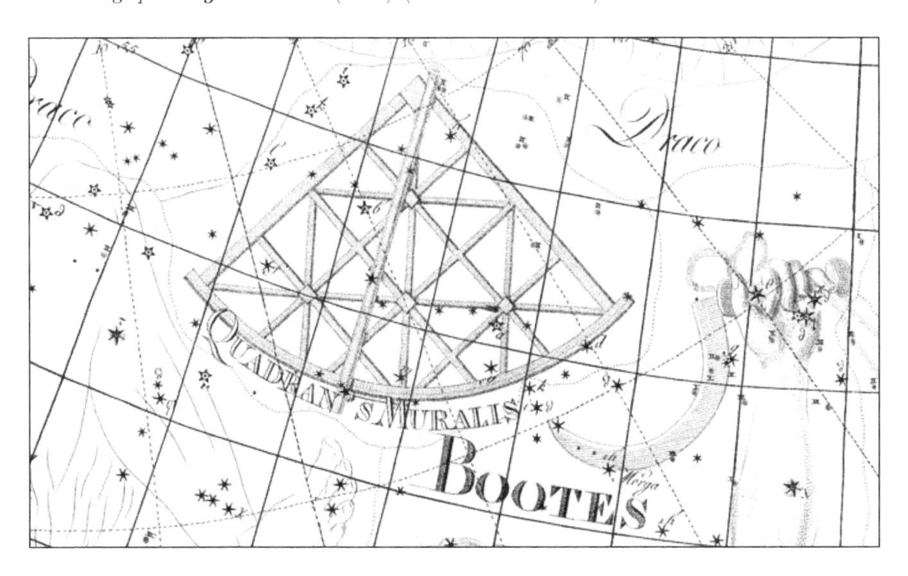

Rangifer

The reindeer

A faint, far-northern constellation introduced in 1743 on a star chart published by the Frenchman Pierre-Charles Le Monnier (1715–99) in his book *La Théorie des Comètes*. The chart showed the track of the comet of 1742 through the north polar region of the sky and Le Monnier was inspired to place a new constellation representing a reindeer on the comet's course, close to the north celestial pole between Cepheus and Camelopardalis. Le Monnier chose a reindeer as a

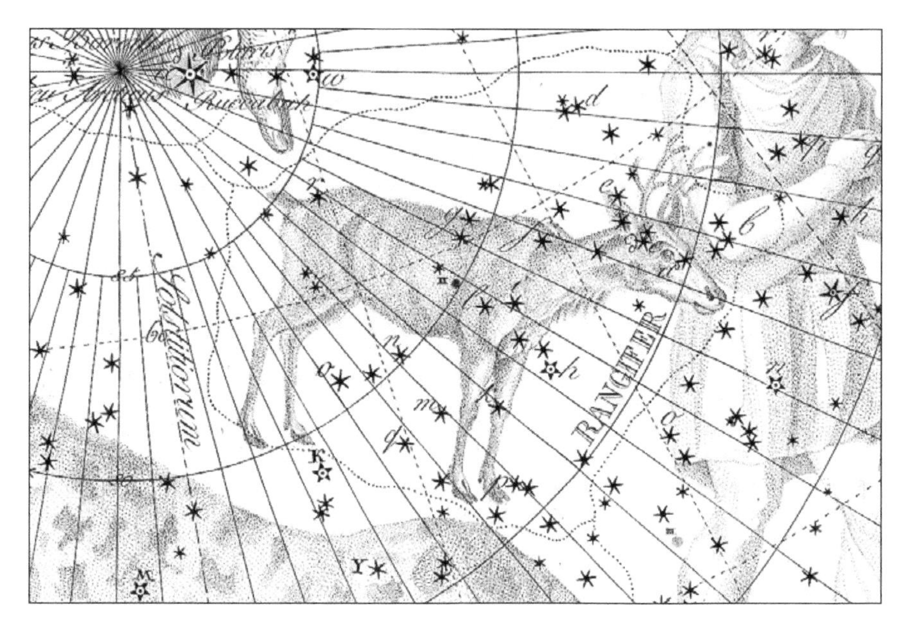

Rangifer seen on Chart III of the *Uranographia* of Johann Bode (1801). Its feet stand almost on the back of Camelopardalis at bottom left, while its head and antlers brush the arm of Custos Messium (page 191), another now-obsolete constellation. The reindeer's tail points towards the north celestial pole, at upper left. (Deutsches Museum)

reminder of his trip to Lapland in 1736–37 with the expedition of Pierre Louis de Maupertuis to measure the length of a degree of latitude in the far north.

On Le Monnier's chart the constellation was named 'le Réene', which should more accurately have been 'le Renne'. Bode Latinized it as Rangifer on his *Uranographia* of 1801. The constellation was sometimes also known as Tarandus from the reindeer's scientific name, *Rangifer tarandus*.

Robur Carolinum

Charles's oak

Edmond Halley (1656–1742) planted this constellation in the southern sky in 1678 as a patriotic gesture to his monarch, Charles II of England. It commemorated the oak tree in which King Charles hid after his defeat by Oliver Cromwell's republican forces at the Battle of Worcester in 1651. Halley formed the constellation from stars that were previously part of Argo Navis.

The invention arose from Halley's visit to the island of St Helena in the south Atlantic Ocean in 1676 to observe the southern sky. He presented his results to the Royal Society in London on his return in 1678 and the following year published his catalogue of southern stars, *Catalogus Stellarum Australium*, with an accompanying map, both of which featured the new constellation.

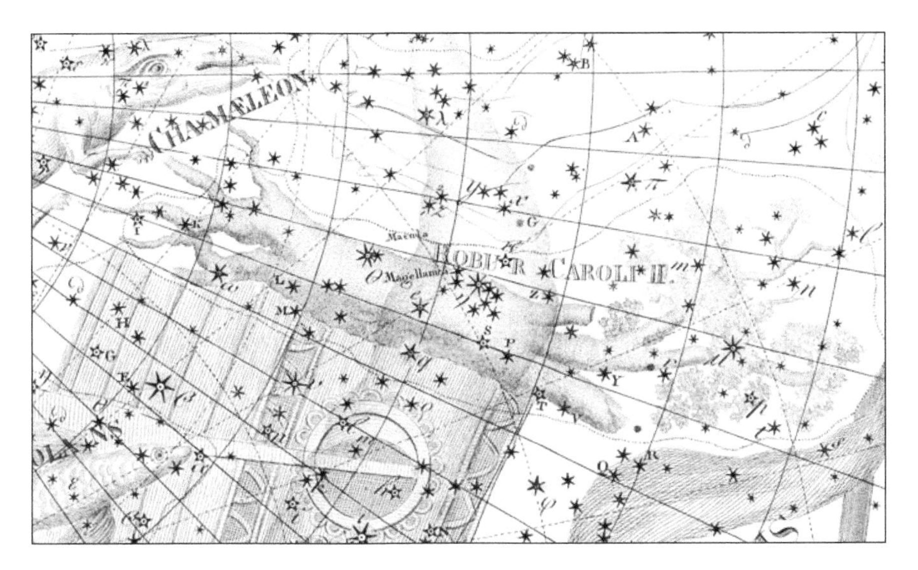

Robur Carolinum shown under the name Robur Caroli II on Chart XX of the *Uranographia* of Johann Bode (1801). It was positioned where the hull of Argo Navis (bottom of picture) was cut off, a place occupied on other maps by either the Clashing Rocks or a bank of clouds. (Deutsches Museum)

Halley described his new constellation as being a 'perpetual memory' of the King, but it turned out to be less permanent than either of them would have hoped. The royal oak was uprooted by the French astronomer Nicolas Louis de Lacaille who mapped the southern stars more comprehensively 75 years after Halley. Most astronomers followed suit in ignoring this genuflection to an English king, although Bode included it on his *Uranographia* atlas of 1801 as Robur Caroli II (see the illustration above).

Sceptrum Brandenburgicum

The Brandenburg sceptre

The German astronomer Gottfried Kirch (1639–1710) introduced this constellation in 1688 to honour the Brandenburg province of Prussia, or more likely its ruler Frederick III. Sceptrum Brandenburgicum lay near the foot of Orion in a large bend in the river Eridanus. In its initial configuration it consisted of a row of five stars of 4th to 6th magnitudes forming a ceremonial sceptre.

The sceptre first appeared on a chart published by Kirch in the scientific journal *Acta Eruditorum* in 1688, but was ignored for nearly a century until Johann Bode, another German, revived it in 1782 in his popular-level star atlas called *Vorstellung der Gestirne*. Bode included it again on his *Uranographia* atlas of 1801 (see overleaf) with extra stars. Even Bode's endorsement, though, could not prevent the eventual extinction of this overtly political creation.

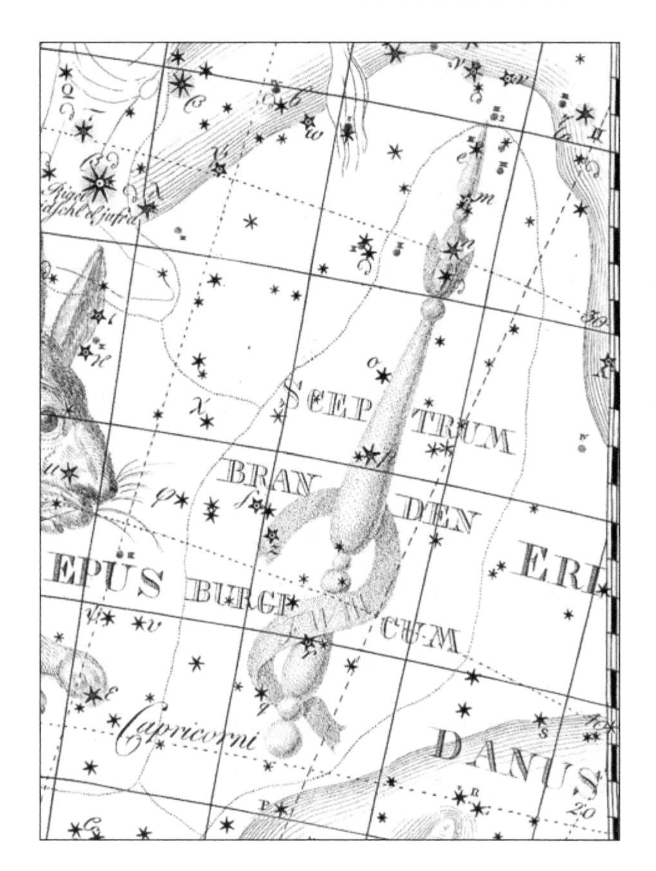

Sceptrum Brandenburgicum lay in a large bend in the river Eridanus, as seen on Chart XVIII of Johann Bode's *Uranographia* star atlas of 1801. Bode inscribed the initials 'FW III' on the ribbon around the lower part of the sceptre in reference to Friedrich Wilhelm III, who had become King of Prussia in 1797. This area of sky has since been returned to Eridanus. (Deutsches Museum)

Taurus Poniatovii

Poniatowski's bull

Everyone has heard of the zodiacal constellation of Taurus the bull, but far less well-known is that there was once a much smaller bull in the sky as well. This little bull, tucked awkwardly between Ophiuchus and Aquila and overlapped by the tail of Serpens, was originated in 1777 by Martin Poczobut (1728–1810), director of the Royal Observatory at Vilna (now Vilnius, Lithuania). Poczobut invented it to honour his king, Stanisław August Poniatowski, who was monarch of both Poland and Lithuania. King Stanisław was a noted patron of the arts and sciences, and the bull was a feature of his family's coat of arms.

The face of the little bull was formed by a V-shaped group of stars between the right shoulder of Ophiuchus and the tail of Serpens. This group reminded Poczobut of the Hyades cluster that outlines the face of Taurus the bull in the zodiac. Four of these had been listed by Ptolemy as 'unformed' stars outside Ophiuchus. A fifth unformed star listed by Ptolemy lay on the bull's right horn and was the brightest of the constellation, magnitude 3.7. The rest of the body was fleshed out by fainter stars. The constellation was first depicted in 1778 as

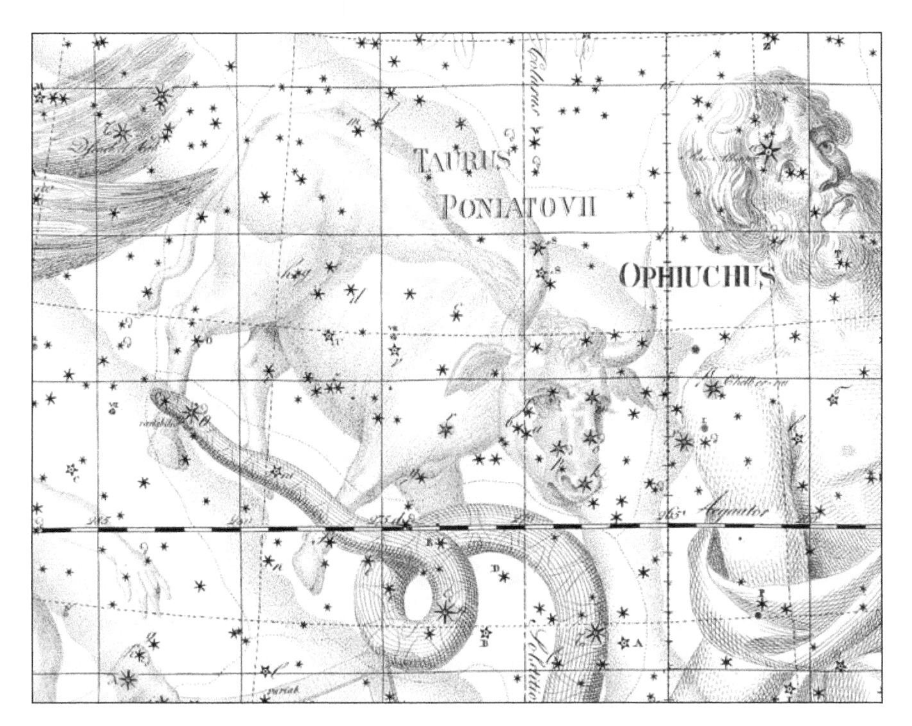

Taurus Poniatovii, lying above the tail of Serpens, the serpent, pictured on Chart IX of the *Uranographia* of Johann Bode (1801). The face of the bull is marked by a V-shaped group of stars reminiscent of the Hyades cluster in Taurus, the zodiacal bull. (Deutsches Museum)

le Taureau Royal de Poniatowski in a revised reprint of Jean Fortin's *Atlas Céleste*. Bode later Latinized its name to Taurus Poniatovii on his *Uranographia* of 1801 (see illustration above).

Poczobut did not realize it but his short-lived creation contained the faint red dwarf known as Barnard's Star, now known to be the second-closest star to the Sun, 5.9 light years away. It lay in the bull's head, near the present-day 66 Ophiuchi. Sadly, Poniatowski's bull did not last long in the sky before being put out to pasture and its stars are now part of Ophiuchus and Serpens Cauda.

Telescopium Herschelii

Herschel's telescope

There were originally two constellations of this name, both introduced in 1789 by the Hungarian-born astronomer Maximilian Hell (1720–92), director of the Vienna Observatory, to commemorate William Herschel's discovery of the planet Uranus eight years earlier. Hell first showed them in a special publication called *Monumenta, Aere Perenniora, Inter Astra Ponenda* which was issued to announce these inventions along with a third constellation of his devising, Psalterium

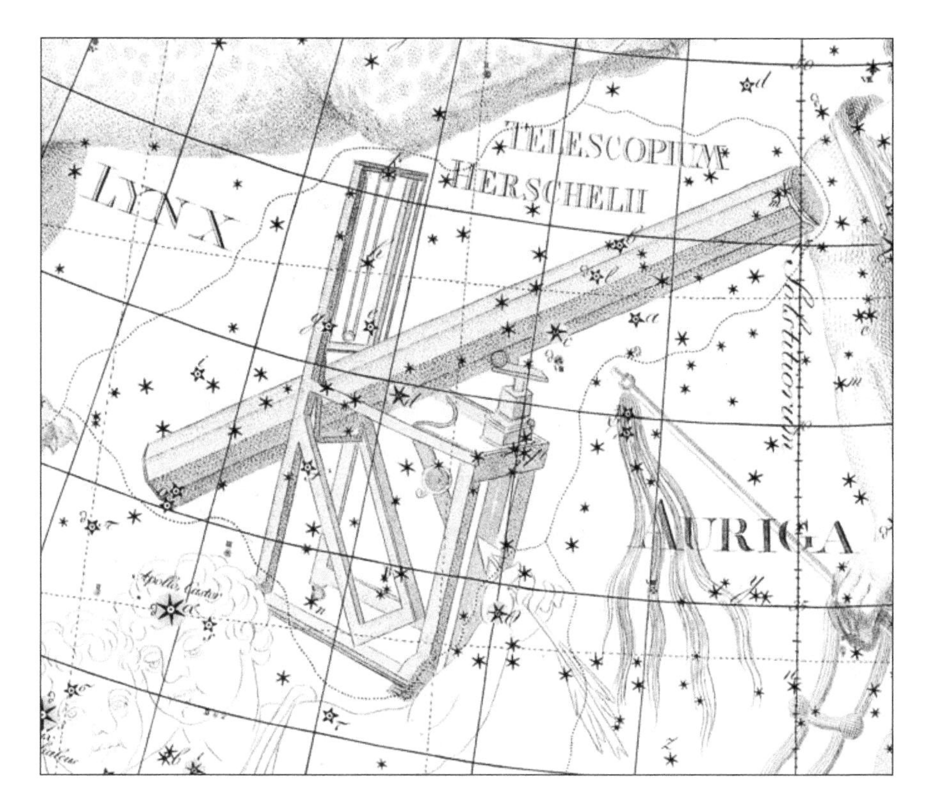

Telescopium Herschelii, depicting the reflecting telescope with which William Herschel discovered the planet Uranus in 1781, seen on Chart V of the *Uranographia* atlas of Johann Bode (1801). The telescope had a wooden tube 7 feet long with a mirror of 6.2 inches diameter. In those days telescopes were described by their length rather than the diameter of their mirror, so Herschel referred to this as his 7-ft telescope. (Deutsches Museum)

Georgianum (see Harpa Georgii, page 194). The two telescopes were positioned either side of the area where the new planet had been found, near the star Zeta Tauri.

Tubus Hershelii Major, as Hell named it, represented Herschel's 20-ft-long telescope and lay between Gemini, Lynx, and Auriga. Tubus Hershelii Minor (again, Hell's spelling), crammed awkwardly between Orion and the head of Taurus, represented Herschel's 7-ft reflector. Judging by the inaccurate representations, though, Hell had not seen either telescope – in the case of Tubus Hershelii Minor he even got the type of telescope wrong, depicting it as a refractor whereas Herschel used only reflectors of his own construction.

Johann Bode reduced the constellations to one in his *Uranographia* atlas of 1801 under the name Telescopium Herschelii, located where Hell had placed Tubus Hershelii Major. Bode, having bought telescopes from Herschel, knew what they looked like and he realistically depicted the 7-ft reflector with which Herschel actually made the discovery of Uranus. Eventually its stars were returned to Auriga, Gemini, and Lynx, from where they had been borrowed.

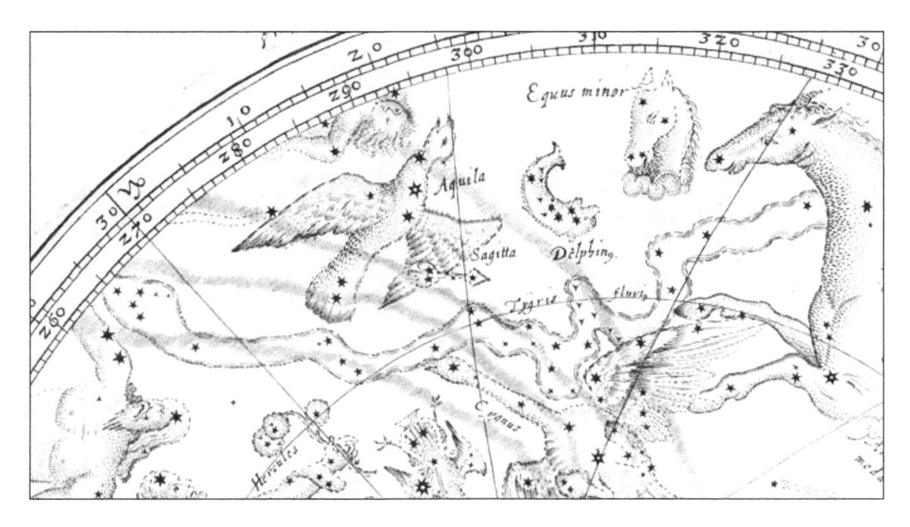

Tigris running from the neck of Pegasus at right, between the head of Cygnus and the tail of Aquila, to the shoulder of Ophiuchus at left, as seen on a planisphere from the 1666 edition of *Planiglobium Coeleste, et Terrestre* by Isaac Habrecht. (ECHO, Berlin.)

Tigris

A constellation representing the river Tigris, a real river of Mesopotamia which joins with the Euphrates in the modern Iraq. The constellation was introduced in 1612 by the Dutchman Petrus Plancius on the same globe that the river Jordan (Jordanus) made its first appearance (see page 196). The celestial Tigris began at the neck of Pegasus and flowed between Cygnus and Aquila, an area now occupied by Johannes Hevelius's later invention Vulpecula. It ended by the right shoulder of Ophiuchus in a V-shaped group of stars that Ptolemy had listed as lying outside Ophiuchus; these were later to be incorporated in another short-lived constellation, Taurus Poniatovii (page 204).

Like Jordanus, Tigris first appeared in print on the 1624 chart by the German astronomer Jacob Bartsch (*c.*1600–33) in his book *Usus Astronomicus Planisphaerii Stellati*. Also like Jordanus, it was not adopted by Hevelius for his influential atlas of 1687. Both constellations became forgotten during the 18th century and were not shown by Johann Bode on his *Uranographia* atlas of 1801.

Triangulum Minus

The lesser triangle

One of the least imaginative constellations, Triangulum Minus was invented in 1687 by Johannes Hevelius. It was formed from three 5th-magnitude stars first catalogued by Hevelius himself. Triangulum Minus lay just south of the existing celestial triangle, Triangulum, which Hevelius renamed Triangulum Majus. The

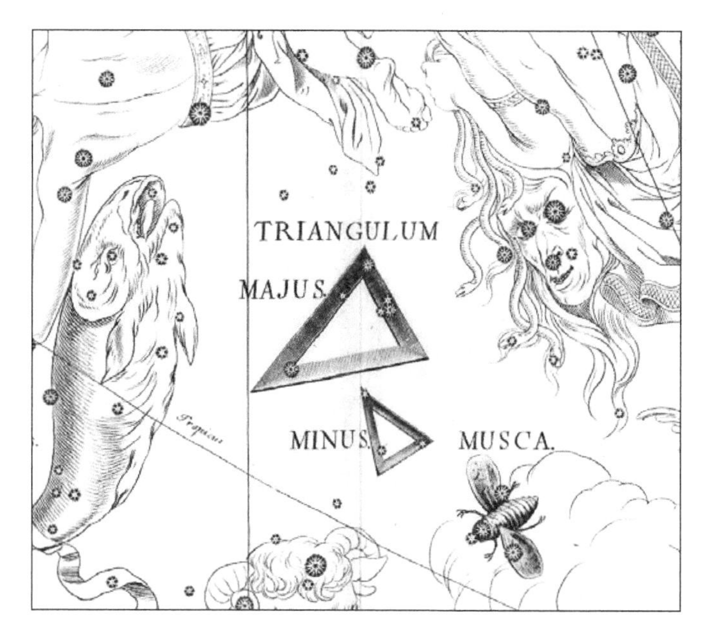

Triangulum Minus shown next to the classical Greek triangle, here renamed Triangulum Majus, on the *Firmamentum Sobiescianum* star atlas of Johannes Hevelius, published posthumously in 1690. Hevelius showed the constellations as they would appear on a celestial globe, so this is a mirror image of how they are seen in the sky; for a view of the two triangles the right way round see page 168. Next to Triangulum Minus is another doomed constellation, Musca Borealis (see page 198). (ETH-Bibliothek Zürich)

little triangle achieved surprisingly wide acceptance among astronomers, including Johann Bode who showed it on his *Uranographia* atlas of 1801. On some maps the pair were known by the combined name Triangula. Ultimately, though, the little triangle was deemed superfluous to requirements when the constellations came to be rationalized.

Turdus Solitarius

The solitaire

This flighty and ultimately doomed constellation was introduced in 1776 by the French astronomer Pierre-Charles Le Monnier (1715–99) in a paper titled Constellation du Solitaire published in the *Mémoires* of the French Royal Academy of Sciences. He described it as a 'bird of the Indies and the Philippines', evidently without realizing that these are two separate birds. It was perched on the end of the tail of Hydra, the water snake, with its head awkwardly overlapping the southern pan of Libra, the scales. Le Monnier said that he introduced the constellation in memory of the voyage to the island of Rodrigues in the Indian Ocean by another French scientist, Alexandre Guy Pingré, who observed the transit of Venus from there in 1761.

Presumably Le Monnier had intended his constellation to represent the Rodrigues solitaire, a flightless bird related to the Dodo which Pingré had hunted for on the island but was unable to find, for they were by then on their way to extinction. However, the bird shown on Le Monnier's diagram of the constellation resembles a female blue rock thrush (*Monticola solitarius* of the family Turdidae), dubbed the 'solitaire of the Philippines' by the French ornithologist Mathurin Jacques Brisson. This is a quite different bird from the Rodrigues solitaire, so it seems that when Le Monnier came to illustrate his new constellation he chose the wrong solitaire. Johann Bode changed its name to Turdus Solitarius in his *Uranographia* atlas of 1801 (see illustration below).

The British scientist Thomas Young (1773–1829) renamed the constellation the Mockingbird on a star chart published in 1807 in *A Course of Lectures on Natural Philosophy and the Mechanical Arts*, while the British amateur astronomer Alexander Jamieson (1782–1850) transformed it into Noctua, the owl, on his *Celestial Atlas* of 1822. But it eventually became as extinct as the bird after which it was originally named.

Turdus Solitarius perched on the tail of Hydra, the water snake, and with its head overlapping the southern pan of Libra, the scales, as depicted on Chart XIV of the *Uranographia* of Johann Bode (1801). (Deutsches Museum)

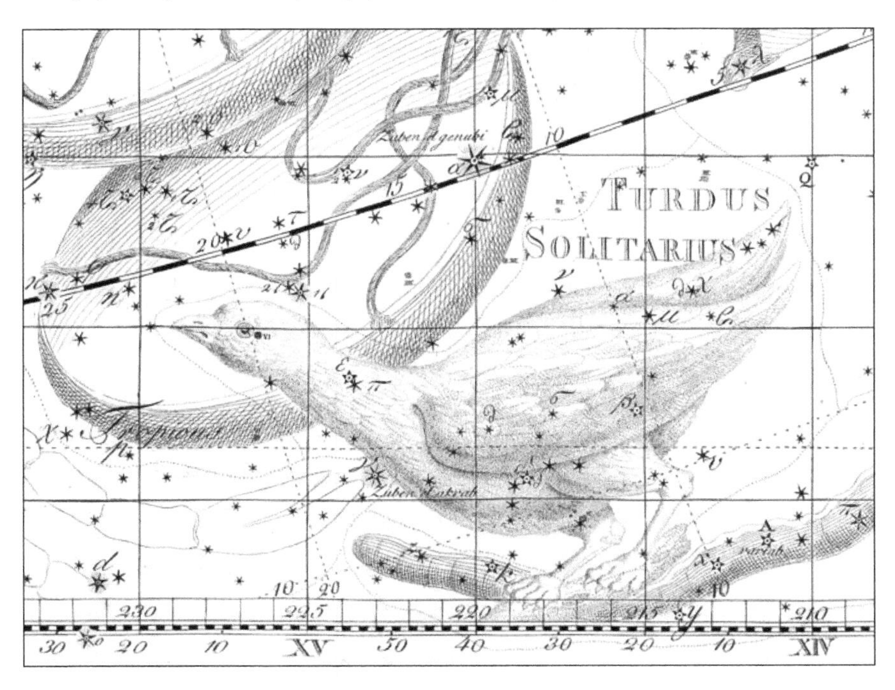

Sources and acknowledgements

For anyone entering the field of Greek mythology, the two volumes by Robert Graves entitled *The Greek Myths* (Penguin) are a masterful synthesis, with copious references. Another useful summary, with many notes and references, is *A Handbook of Greek Mythology* by H.J. Rose (Methuen). For other background information I consulted the *Oxford Classical Dictionary* (Oxford University Press) and the *Dictionary of Classical Mythology* by Pierre Grimal (Blackwell); the latter, in particular, contains a fund of references.

The starting point for all studies of Greek star lore is a poem called the *Phaenomena* (Appearances), written *c.*275 BC by Aratus of Soli. The *Phaenomena* of Aratus is based on a book of the same name written the previous century by the Greek scientist Eudoxus of Cnidus. No copies of the book by Eudoxus have been preserved; we have only Aratus's poem. An English translation by G. R. Mair is available in the Loeb Classical Library series (Harvard University Press and Heinemann). A more recent translation and extensive commentary on the poem is *Aratus: Phaenomena* by Douglas Kidd (CUP, 1997).

The Latin adaptation of Aratus that was reputedly written by Germanicus Caesar in the early part of the first century has been translated by D. B. Gain; see *The Aratus Ascribed to Germanicus Caesar* (Athlone Press, 1976). A Latin work with many echoes of Aratus is *Astronomica* by the Roman poet Marcus Manilius, written early in the first century AD. It has been translated into English by G. P. Goold in the Loeb Classical Library.

Another early Greek source is the *Catasterisms* ascribed to Eratosthenes in the second century BC – although not, according to modern authority, actually written by him. When writing the first edition of this book I could find no record of an English translation; instead, I referred to the French version published in 1821 by Abbé Halma. Since then, English translations have appeared in the books *Star Myths* by Theony Condos (Phanes Press, Grand Rapids, 1997) and *Constellation Myths* by Robin Hard (OUP, 2015).

The Myths of Hyginus by Mary Grant (University of Kansas Publications, 1960) contains an invaluable English translation of Hyginus's *Fabulae* and *Poetic Astronomy*, among the most influential works on constellation mythology but scarcely read today. Other translations of the *Poetic Astronomy* are included in the more recent *Star Myths* by Theony Condos and *Constellation Myths* by Robin Hard, mentioned above.

Apollodorus was a Greek writer who produced an encyclopedic summary of Greek myths called the *Library*; I referred to the Loeb translation by Sir J. G. Frazer. Many popular myths received their definitive retelling in the works of the Roman writer Ovid; for his *Metamorphoses* I used the Penguin translation by Mary Innes and the Loeb edition of his *Fasti* by Sir J. G. Frazer. My source for Apollonius Rhodius was the Penguin translation by E. V. Rieu.

For Ptolemy's *Almagest* I consulted G.J. Toomer's thoughtful translation (Duckworth, 1984).

A Greek writer called Geminus (probably first century BC) gives us a glimpse of the Greek sky between the eras of Hipparchus and Ptolemy in his *Introduction to the Phenomena* (commonly known as the *Isagoge*, from the first word in its Greek title *Eisagoge eis ta phainomena*). The first English translation of this book was published in 2006 by James Evans and J. Lennart Berggren.

For the origin of star names, I have relied on the booklet *A Dictionary of Modern Star Names* by Paul Kunitzsch and Tim Smart (Sky Publishing, 2006). Useful background on star names can also be found in an article by Paul Kunitzsch in the January 1983 issue of *Sky & Telescope*. A collection of Kunitzsch's scientific papers has been reprinted as *The Arabs and the Stars* (Variorum, 1989). An illuminating paper by Gwyneth Heuter on the origin of star names is to be found in *Vistas in Astronomy*, vol. 29, 1986, p. 237.

The Sky Explored by Deborah Jean Warner (Alan R. Liss, New York, and Theatrum Orbis Terrarum, Amsterdam, 1979) is an invaluable survey of the history and development of celestial cartography, and contains much incidental material on constellation history. A notable work on the modern constellations, *Filling the Sky* published in 2003 by Jim Fuchs, provided useful additional information and references when I came to revise the text for this second edition. Morton Wagman's *Lost Stars* (McDonald & Woodward) is a painstaking survey of changes in star designations between catalogues and in response to changing constellation boundaries, as well as being a good survey of constellation history.

Archie Roy's speculations about the origin of the constellations are contained in his paper in *Vistas in Astronomy*, vol. 27, 1984, p. 171. Arguments by Bradley E. Schaefer for a later date of origin are to be found in *Journal for the History of Astronomy*, vol. 35, 2004, p. 161. E. B. Knobel's analysis of the star catalogue of Frederick de Houtman is in the *Monthly Notices of the Royal Astronomical Society*, vol. 77, 1917, p. 414.

R. H. Allen's *Star Names, Their Lore and Meaning* (Dover) and W. T. Olcott's *Star Lore of All Ages* (Putnam's) are fun to dip into, but I have not used them as prime sources for mythology.

For this second edition I have replaced all illustrations with improved versions. My thanks go to the various libraries and institutions around the world who have made high-quality scans of various historical celestial atlases and catalogues freely available; the sources are credited individually at each illustration. In addition I wish to thank Ilia A. Shurygin of St Petersburg, Russia, for allowing me to use his excellent image of the Farnese Atlas.

References

Baily, Francis, *The Catalogues of Ptolemy, Ulugh Beigh, Tycho Brahe, Halley, Hevelius..., Memoirs of the Royal Astronomical Society*, vol. 13, 1843.

Condos, Theony, *Star Myths* (Phanes Press, Grand Rapids), 1997.

Dekker, Elly, *Der Globusfreund*, nos. 35–37, pp. 211–30, 1987.

Dekker, Elly, *Annals of Science*, vol. 44, pp. 439–70, 1987.

Dekker, Elly, *Annals of Science*, vol. 47, pp. 529–60, 1990.

Evans, David S., *Lacaille: Astronomer, Traveler* (Pachart, Tucson), 1992.

Evans, James, and Berggren, J. Lennart, *Geminos's Introduction to the Phenomena* (Princeton University Press), 2006.

Frazer, J. G., *Apollodorus* (2 vols: Loeb Classical Library), 1921.

Frazer, J. G., *Ovid's Fasti* (Loeb Classical Library), 1931.

Fuchs, Jim, *Filling the Sky* (privately published), 2003.

Gain, D. B., *The Aratus Ascribed to Germanicus Caesar* (Athlone Press, London), 1976.

Goold, G. P., *Manilius Astronomica* (Loeb Classical Library), 1977.

Grant, Mary, *The Myths of Hyginus* (University of Kansas Publications, Lawrence), 1960.

Graves, Robert, *The Greek Myths* (2 vols: Penguin), 1960.

Grimal, Pierre, *The Dictionary of Classical Mythology* (Blackwell, Oxford), 1985.

Hard, Robin, *Constellation Myths* (Oxford University Press), 2015.

Heuter, Gwyneth, *Vistas in Astronomy*, vol. 29, p. 237, 1986.

Innes, Mary M., *Ovid Metamorphoses* (Penguin), 1955.

Kidd, Douglas, *Aratus: Phaenomena* (Cambridge University Press), 1997.

Knobel, E. B., *Monthly Notices of the Royal Astronomical Society*, vol. 77, p. 414, 1917.

Kunitzsch, Paul, *Sky & Telescope*, vol. 65, p. 20, 1983.

Kunitzsch, Paul, *The Arabs and the Stars* (Variorum, Northampton), 1989.

Kunitzsch, Paul, and Smart, Tim, *A Dictionary of Modern Star Names* (Sky Publishing), 2006.

Mair, G. R., *Aratus* (Loeb Classical Library), 1955.

Oxford Classical Dictionary (second and third edns: Oxford University Press), 1970, 1999.

Rieu, G. V., *Apollonius of Rhodes The Voyage of Argo* (Penguin, London), 1971.

Rose, H. J., *A Handbook of Greek Mythology* (Methuen, London), 1958.

Roy, Archie, *Vistas in Astronomy*, vol. 27, p. 171, 1984.

Schaefer, Bradley E., *Journal for the History of Astronomy*, vol. 35, p. 161, 2004.

Toomer, G. J., *Ptolemy's Almagest* (Duckworth, London), 1984.

Volkoff, Ivan, et al., *Johannes Hevelius and his Catalog of Stars* (Brigham Young University Press, Provo), 1971.

Wagman, Morton, *Lost Stars* (McDonald & Woodward, Blacksburg), 2003.

Warner, Deborah Jean, *The Sky Explored* (Alan R. Liss, New York, and Theatrum Orbis Terrarum, Amsterdam), 1979.

Wender, Dorothea S., *Hesiod and Theognis* (Penguin), 1973.

Glossary of mythological characters

The Greeks and Romans had similar gods and mythological characters but used different names for them. Hence what may sound at first to be two separate individuals, such as Zeus and Jupiter, are really one and the same. This table lists the Latin equivalents (*italicized* names) of the major Greek characters mentioned in this book.

Greek name	Latin name	Greek name	Latin name
Aphrodite	*Venus*	Ares	*Mars*
Artemis	*Diana*	Asclepius	*Aesculapius*
Athene	*Minerva*	Cronos	*Saturn*
Demeter	*Ceres*	Dionysus	*Bacchus*
Eros	*Cupid*	Hades	*Pluto*
Hephaestus	*Vulcan*	Hera	*Juno*
Heracles	*Hercules*	Hermes	*Mercury*
Persephone	*Proserpina*	Polydeuces	*Pollux*
Poseidon	*Neptune*	Zeus	*Jupiter*

Index

Page numbers in **bold** type refer to illustrations.

About the author

Ian Ridpath is an English writer and editor on astronomy and space. He is author of a standard series of observing guides for amateur astronomers: the *Collins Stars & Planets Guide* (known in the US as the *Princeton Stars & Planets Field Guide*); Collins Gem *Stars*; and *The Monthly Sky Guide*, all illustrated by Wil Tirion, the world's foremost celestial cartographer. He is editor of the authoritative *Oxford Dictionary of Astronomy* and of the last three editions of *Norton's Star Atlas*, the longest-established and best-known star atlas in the world. Ian is a recipient of the Astronomical Society of the Pacific's Klumpke-Roberts Award for 'outstanding contributions to the public understanding and appreciation of astronomy', the most prestigious award of its kind. He is a Fellow of the Royal Astronomical Society and a member of the International Astronomical Union, where he serves on its Working Group on Star Names. His interests include collecting antique astronomy books, particularly star atlases.

www.ianridpath.com

BV - #0147 - 131123 - C0 - 234/156/14 - PB - 9780718894788 - Gloss Lamination